Virtually any exercise in ecology will require some knowledge of the techniques for carrying out a census of population numbers. This practical text outlines clearly, with worked examples, the main techniques used by field ecologists to enumerate plants and animals. Each taxonomic group is treated separately, with detailed descriptions of appropriate census methods, their advantages, disadvantages and biases. Techniques for measuring a wide range of environmental variables are also included. The final chapter lists the 20 most common censusing sins.

Concise yet comprehensive, this book provides a unique overview of the most important methods for those working on field studies in population and behavioural ecology and conservation biology at all levels, from the beginner to the practising professional.

Ecological Census Techniques
a handbook

Ecological Census Techniques
a handbook

Edited by

WILLIAM J. SUTHERLAND
University of East Anglia

CAMBRIDGE
UNIVERSITY PRESS

Published by the Press Syndicate of the University of Cambridge
The Pitt Building, Trumpington Street, Cambridge CB2 1RP
40 West 20th Street, New York, 10011-4211, USA
10 Stamford Road, Oakleigh, Melbourne 3166, Australia

First published 1996

Printed in Great Britain by The Bath Press, Avon

A catalogue record for this book is available from the British Library

Library of Congress cataloguing in publication data

Ecological census techniques: a handbook / William J. Sutherland (ed.).
  p. cm.
  Includes bibliographical references and index.
  ISBN 0 521 47244 X (hb). – ISBN 0 521 47815 4 (pb)
  1. Ecological survey – methodology. I. Sutherland, William J.
QH541.15.S95E26 1996
574.5'028 – dc20 95–31985 CIP

ISBN 0 521 47244 X hardback
ISBN 0 521 47815 4 paperback

Contents

4 Invertebrates 139
MALCOLM AUSDEN

5 Fish 178
MARTIN R. PERROW, ISABELLE M. CÔTÉ, AND MICHAEL EVANS

11   The twenty commonest censusing sins   317
WILLIAM J. SUTHERLAND

Contributors

MALCOLM AUSDEN
School of Biological Sciences, University of East Anglia, Norwich NR4 7TJ, United Kindom

SIMON BLOMBERG
Department of Anatomical Sciences, University of Queensland, St Lucia, Queensland 4072, Australia

JAMES BULLOCK
Furzebrook Research Station, NERC Institute of Terrestrial Ecology, Wareham, Dorset BH20 5AS, United Kingdom

ISABELLE M. CÔTÉ
School of Biological Sciences, University of East Anglia, Norwich NR4 7TJ, United Kingdom

MICHAEL EVANS
National Rivers Authority, Anglian Region, 79 Thorpe Road, Norwich NR1 1EW, United Kingdom

DAVID W. GIBBONS
The Royal Society for the Protection of Birds, The Lodge, Sandy, Bedfordshire SG19 2DL, United Kingdom

JEREMY J. D. GREENWOOD
British Trust for Ornithology, Thetford, Norfolk IP24 2PU, United Kingdom

TIMOTHY R. HALLIDAY
Department of Biology, The Open University, Milton Keynes MK7 6AA, United Kingdom

DAVID HILL
Ecoscope Applied Ecologists, 9 Bennell Court, Comberton, Cambridge CB3 7DS, United Kingdom

JACQUELYN C. JONES
School of Biological Sciences, University of East Anglia, Norwich NR4 7TJ, United Kingdom

MARTIN R. PERROW
    ECON Ecological Consultancy, School of Biological Sciences, University of East Anglia, Norwich NR4 7TJ, United Kingdom

JOHN D. REYNOLDS
    School of Biological Sciences, University of East Anglia, Norwich NR4 7TJ, United Kingdom

RICHARD SHINE
    Zoology A08, School of Biological Sciences, University of Sydney, Sydney NSW 2006, Australia

WILLIAM J. SUTHERLAND
    School of Biological Sciences, University of East Anglia, Norwich NR4 7TJ, United Kingdom

Preface

Almost all ecological and conservation work requires censusing, and the aim of this book is to describe the main techniques. I asked the authors of each chapter to outline those techniques which are most widely used for that taxonomic group and state the problems and advantages associated with each.

I thank the students on the Tropical Biological Association field course at Kibale Forest, Uganda and undergraduates at the University of East Anglia for testing and commenting on an earlier draft of this book.

University of East Anglia   William J. Sutherland

1 Why census?

William J. Sutherland

School of Biological Sciences, University of East Anglia, Norwich NR4 7TJ,
United Kingdom

Table 1.1. *Contents of Chapter 1*

Introduction

The aim of this book is to outline the main techniques for ecological censusing. There are many reasons for deciding to carry out a census. It may be to determine the importance of a site, the population size of a species, the habitat requirements of a species, the reasons for the species' decline, or whether habitat management has been a success, or to understand the population dynamics. In this chapter I will outline the different approaches necessary for tackling each of these questions and describe some common problems and mistakes.

It is important to plan the work carefully. Many studies are overambitious and thus waste time and effort and often fail. For example, it is a common mistake to collect far more samples than can ever be analysed.

The data must be stored in a way that can be retrieved and understood by others in the future. Notebooks are rarely sufficient and it is usually important that data are stored elsewhere in a form that can be readily interpreted, such as data sheets in files or computer records. It is often useful to document the exact locations where species of particular interest were found.

Describing the interest of sites

One of the major priorities for conservationists is to determine the most important areas for a given species or group. Questions may include: How important is a particular country?

Which are the major areas within a country? Which sites within an area are most important?

Often a major objective is to describe the importance of a site for a range of species. The objective then is perhaps to provide a species list with only a rough assessment of abundance. In this case use as wide a range of different techniques as possible. Thus sample in as many habitats, over as wide a period of time as is practical. Especially for invertebrates, it is best to sample at different times of day, on different dates if possible, under different weather conditions and using different modifications of the same techniques. Even minor modifications, such as changing the preservative in pitfall traps or the colour of a water trap, will influence the species caught.

When preparing species lists for an area it is often useful to draw a species incidence curve. This plots cumulative total number of species seen against number of days spent searching. This gives a good indication as to whether further searching will increase the number of species recorded.

Box 1.1. **Determining the ornithological importance of Gola Forest**

The Upper Guinea forest block in West Africa holds 22 endemic and 8 globally threatened bird species. The Gola Forest Reserves in south-east Sierra Leone are among the last fragments of this formerly extensive habitat, yet they had been visited by only one ornithologist, and only 3 of the threatened species had been positively identified.

In 1988/9 a team of surveyors set out to determine the importance of the Gola Forest Reserves. A major objective was locating the most important species and this entailed using a range of techniques. Bees' nests were watched for feeding Yellow-footed Honeyguide *Melignomon eisentrauti* and the canopy was searched for this as well as Nimba Flycatcher *Melaenornis annamarulae* and Western Wattled Cuckoo-shrike *Campephaga lobata*. White-breasted Guineafowl *Agelastes meleagrides* was detected from its calls and from the scratching noise it made when foraging in litter on the forest floor. Stretches of rivers were watched at dusk for Rufous Fishing Owl *Scotopelia ussheri* and trees beside watercourses were searched during the day for roosting birds. Parties of birds were searched for Yellow-throated Olive Greenbul *Criniger olivaceus*, Spot-winged Greenbul *Phyllastrephus leucolepis* and Gola Malimbe *Malimbus ballmanni*. Rocky areas and cliffs were searched for the distinctive swallow-like mud cup nests of White-necked Picathartes *Picathartes gymnocephalus*. Other techniques used included watching fruiting and flowering trees, checking ant columns for associated birds, imitating bird calls, and using tapes of bird calls to attract birds (Allport *et al*. 1989).

This project was successful in recording 274 species, locating 7 globally threatened species and glimpsing a sunbird almost certainly new to science. The resulting recognition of the importance of this site led to the Gola Rain Forest Conservation Programme.

If the aim is to describe the distribution of a species it is important also to document the area searched. It is essential to be able to tease apart the distribution of the species from the distribution of observers.

Estimating population size

A common objective is to estimate the numbers in an area or even the size of the whole population.

One approach is to try to count all individuals in the entire area. However, this is often impractical, so it is necessary to select sample sites to visit. One common mistake is just to visit the known sites or what is considered to be the best habitat for the species, for example, by visiting all the remaining areas of primary forest but ignoring all the secondary forest. This is only acceptable if it is absolutely certain that the species is restricted to primary forest. It may be that the secondary forest does hold a lower density but also has a greater area and thus actually contains more individuals. Without coverage of all suitable habitats it is impossible to estimate the total population.

The best approach is usually random sampling, since this overcomes such biases and should result in representative coverage. Divide the area into blocks. These could, for example, be 1-km squares or natural divisions, such as lakes or bogs. The more blocks that can be visited the greater the accuracy.

Exactly the same technique is used in sampling with quadrats or soil cores, although the area sampled is likely to be a minute fraction of the total area. The area can be considered as a grid with potential sampling points at the intersections of the grid lines. Thus the study area may be 150 m long and 500 m wide. Draw pairs of random numbers between 0 and 150 and between 0 and 500 and sample at the intersection of these random numbers. Most study areas will not, however, be neat rectangles. In this case overlay a grid that is larger than the study area but ignore those pairs of points that lie outside the study area (see Box 2.21).

The blocks should be numbered and the ones to be surveyed selected using random numbers. Guessing random numbers, however, has been shown to be highly non-random. Random numbers are best determined using random-number tables (one is given in Table 2.6) or a calculator random-number function. However, they can also be improvised by a number of mechanisms, such as using the last digits from a telephone directory, from randomly stopping digital watches and reading the number of hundredths of a second or by guessing a four digit number (e.g. 7217), adding the digits together (17), and repeating until reduced to a single digit (8).

If there are obvious differences in habitat within the survey area then a useful approach is stratified random sampling. This involves dividing the area into different habitats (e.g. mangrove forest, mudflats, and sand banks) and then randomly sampling areas within each. It is then possible to provide an estimate of the number in each habitat and thus the total population size.

Box 1.2. **Surveying the Fuerteventura Stonechat** *Saxicola dacotiae*

Bibby and Hill (1987) set out to estimate the total population of this species and its habitat requirements. Twenty-one blocks, each of twelve 1-km² grid squares (or a smaller area if the block overlapped the sea), were selected at random. Each of the twelve 1-km squares in each block was visited once for about 2–3 hours, which was thought to be sufficient to detect all individuals of this conspicuous species. The habitat in each square was described.

From this survey of 12.7% of the island it was possible to extrapolate to estimate the total population at 750 ± 100 pairs. As the methods and areas surveyed are described precisely, it would be possible to repeat exactly the same survey to look for changes in the population.

Monitoring population changes

Many studies attempt to determine whether and to what extent populations are changing. This may be to monitor the fate of a species of conservation interest or to see whether a pest species is increasing. It is often not necessary to have an absolute population estimate and a relative measure of abundance may be sufficient.

It is essential to ensure that exactly the same techniques of monitoring are used each time. It is very tempting to wander from a transect line to include particularly good patches, or to improve a technique in a way that results in encountering more individuals, but this, of course, makes comparisons with previous years invalid. If the decision is made to change techniques, then it is necessary to have a period of overlap in which both methods are used so the relative efficiency of each can be determined and the data can be calibrated.

It is very useful to monitor environmental variables in such a way that site managers can detect changes and so relate them to changes in populations. For example, if the salinity and water level on a coastal lagoon are recorded on a regular basis, then a warden can tell whether a particular measurement is within the typical range of fluctuations or an indication of an atypical event.

In describing methods for monitoring it is essential to be extremely specific about the exact techniques used so that they can be repeated by another observer in the future. Small and sensible modifications, such a not censusing if the wind is above a certain speed, must be specified so that future observers collect comparable data.

How regularly the censuses need to be made will depend upon the likely rate of change. Censusing trees each decade may be sufficient, while phytoplankton may vary from week to week. The date of the census should ideally be standardised, but if the season is particularly early or late, then it may be sensible to alter the census accordingly.

Box 1.3. **Determining ecological changes in Breckland heaths**

Breckland is an area of early successional lowland grass heath in eastern England of considerable conservation interest. To document the changes in plants, birds and insects four main techniques were used (Dolman and Sutherland 1991, 1992). Each involved locating and collating past records and comparing them with the current situation.

(1) The location of sites of the rare plant species was well documented although there was little information on abundance. Resurveying these sites showed that, for the different species, between 29 to 93% of these populations had disappeared and that in the sites in which they had disappeared, much of the loss was attributable to succession to rank grasslands. (2) On nature reserves there were a few sites with annual bird counts, a few complete surveys, including one in 1949 and one in 1968, and scattered records for a large number of sites extracted from a wide array of sources. These data could be compared with current numbers which showed that many of the characteristic species had since disappeared (Woodlark *Lullula arborea*, Ringed Plover *Charadrius hiaticula*) or plummeted in numbers (Wheatear *Oenanthe oenanthe*, Stone Curlew *Burhinus oedicnemus*). (3) A series of quadrats had been carried out in 1981; they were then resurveyed and this showed a change from the characteristic community of lichens and annual plants towards competitive grasses, shrubs and trees. (4) It was hard to find data for determining any changes in invertebrate abundance. Moths were the best documented and a number of characteristic species had gone extinct. This study thus showed that there had been considerable loss of conservation interest even on nature reserves and it seemed this could be linked to the lack of habitat disturbance by humans and rabbits. Such disturbance is essential for maintaining this habitat.

Determining the habitat requirements of a species

One common and useful objective is to try and determine the habitat requirements of a particular species. This work may be relatively easy and a lot of further studies of this sort are needed. For such studies it is not necessary to have absolute population estimates and relative numbers are often almost as good.

A frequent mistake is just to visit all the best areas in which the species is found and measure various factors relating to the habitat. Without any comparison of sites in which the species is missing, this is of little use. One useful technique is to compare the points at which the species is located with a random collection of points. In such a comparison it may be possible to include prey abundance, predator abundance and nesting sites, habitat structure and environmental variables. Instead of comparing presence and absence, an alternative is to relate relative density to habitat variables, but it is again essential to visit the poorer sites to determine why they are poor.

An alternative approach makes use of the possibility that the area may be readily divisible into ponds, fields or forest blocks. The analysis can then be either to compare blocks which contain the species with those without or to see what, if any, obvious differences there are. For example, a survey of gardens, using a simple questionnaire, showed that snails *Cepea nemoralis* were abundant in gardens with cats (which scare off the predatory Song Thrushes *Turdus philomelus*) but infrequent in gardens with dogs – which chase off the cats. Such studies comparing sites may be carried out on a range of scales using slightly different techniques. For example, compare the woods in which a species occurs with those in which it does not; within a wood compare the locations in which it is observed feeding with random locations, and the trees it breeds in with a random sample of those in which it does not.

Especially for studies of invertebrates, it may be necessary to compare the fine details of the habitat. For example, various different studies of different species of butterflies have shown that females may only lay eggs on sunny leaves, plants adjacent to bare ground, large leaves or plants rooted in deep soil.

Such studies are most useful when combined with research into the natural history of the species. Where possible, studying the diet by means of observations or analysis of prey-remains in faeces is a very useful step.

Determining why species have declined

Many species of conservation interest are declining and understanding why is an important first step in maintaining and enhancing the population. One useful approach is to determine the requirements of a species as described earlier and try and determine whether the factors meeting these requirements have changed. Another very useful approach is to compare the sites in which the species still occurs with the sites from which it has disappeared.

A further approach is to estimate the life history parameters such as fecundity, survival of the early stages, or adult survival. Theoretically, it is then possible to see if the population is capable of sustaining itself, but in practice it is often difficult to measure these parameters with sufficient accuracy. Such measurements may, however, succeed in indicating problems. For example, in many plant species the mature plants survive well, are pollinated and carry reasonable numbers of seeds but fail to germinate successfully. Often in these cases disturbing the ground is necessary for the species to flourish (of course, many plants are largely dependent on clonal growth and in these species seedling recruitment is less important). Alternatively seedling survival may be high but seed set is low owing to grazing. In this case a reduction in livestock is the solution.

Another useful approach is to assess the limiting factors – the factors which reduce the breeding output and survival. Is the breeding success low because the nests were flooded, because few eggs hatched, because insufficient food was brought to the young who then starved or because many young were eaten by predators? Such knowledge leads to suggestions for conservation measures. One approach is to determine whether these limiting factors have changed. Another is to compare the limiting factors in different areas and relate this to population size or population trends.

There are pitfalls with this approach. On a number of occasions studies of rare plants have shown that many of the flowers are grazed which has led to the recommendation that stock is excluded. This, however, often has disastrous consequences for the rare plant, since, although flowering and seed set is initially improved, without the bare ground produced by grazing, little regeneration takes place. Similarly, fire initially seems disasterous, but for many communities such as the American pairies and the woodlands of northern Australia it is essential for the persistance of many species. This pitfall is best overcome either by detailed research into the habitat requirements or by management experiments.

Box 1.4. **Why has the Silver-spotted Skipper butterfly declined?**

Records showed that the Silver-spotted Skipper butterfly *Hesperia comma* had disappeared from many sites on semi-natural chalk grassland in Britain including a quarter of the populations that once bred on nature reserves. Detailed observations showed that the females laid their eggs on the grass *Festuca ovina* especially adjacent to bare areas (Thomas *et al.* 1986). A comparison of sites showed that the population persisted in sites with stock grazing (which creates gaps) but had gone extinct in those without. The decline could then be related to a reduction in grazing and reintroducing stock grazing has proved successful in enhancing population size.

A subsequent experiment showed that, surprisingly, the population was lower in blocks with summer grazing than in those without. However, observations showed that females tended not to lay eggs on grazed leaves. Thus the ideal management for this species seems to be heavy winter grazing to create bare patches followed by removal of stock grazing to allow the grass to regrow.

Monitoring habitat management

A major failing of nature reserve management is that little has been learnt from the huge amount of management that has been carried out. When experiments have been carried out the results may be surprising. For example, conservationists sometimes control the biennial weed, Marsh Ragwort *Senecio aquaticus*, by pulling it up by hand, but when this was analysed in an experiment it turned out that this practice actually increased the ragwort population, probably by disturbing the ground (Ausden 1993). Such experiments can be used to tackle a range of subjects: Which grazing regime is the best? Do human visitors affect the population? Does harvesting have long-term consequences?

The usual non-experimental approach is to notice that two aspects change in parallel (e.g. there has been an increase in number of visitors to an area and at the same time a decrease in a species). From this it is argued that the increase in visitors is the cause of the decline. Although this may be the case, it may also be that the species has changed for other reasons, such as a change in habitat, and there are innumerable examples of such faulty reasoning. A properly designed experiment overcomes this problem.

An experiment should be controlled, randomised and replicated. Controls are areas in which the management has not been altered. Thus an experiment may compare areas that have been flooded with those that have not. The control means it is possible to separate population changes, for example due to the weather, from the consequences of management.

The treatments (for example flooding versus not flooding) need to be allocated at random. One common failing is to assume that the management will be beneficial and thus manage the best areas. It is then of course difficult to make comparisons and this may give the impression that the management has been successful (as there are more individuals in the managed areas) when it was not. There may be good reasons for managing part of a site in a particular way. Rather than pretending this is part of a randomised design, it is better to exclude such areas from the experiments, divide the remaining area into blocks, and then genuinely randomise the treatments.

Experiments need to be large enough for sufficient individuals to occupy the treatments. If only a few individuals occur in the area then any change could be attributed to chance. For example, if the total population increases in the managed areas from 4 to 6 but stays at 4 in the control areas, it is difficult to determine if this is due to chance.

Replication is important as any two sites are bound to differ anyway. By using say ten blocks and flooding half at random it is then possible to see if the pattern is consistent.

Contrary to widespread belief, it is not essential to carry out a census before the management occurs. With a suitable randomised, replicated, and controlled experiment the effect of treatment is determined by the simple comparison of the managed areas with the controls.

Another useful approach is to pick a number of sites, such as fields, and manipulate part of each. Again it is essential to select randomly which part is manipulated and which is the control. Thus an experiment may consist of five fields, in which half of each field is cattle grazed and the other half is sheep grazed. This is a useful experimental design when the sites differ.

Box 1.5. **When should prairies be burnt?**

Fire is known to be essential in maintaining tallgrass prairies in North America. Natural fires are usually caused by lightning and occur in late summer, while conservationists usually burn prairies before the growing season. Does this matter? Howe (1994) divided an experimental prairie into 21 plots, each of 12 m by 15 m, and burnt 7 on 31 March and 7 on 15 July, while a further 7 were left unburnt. The vegetation was recorded in 12 randomly placed 1-m^2 quadrats within each of the 21 plots. The results showed that late burns resulted in a different community of greater diversity which consisted of smaller and early flowering species.

Population dynamics

By measuring life history parameters and estimating the population size it is possible to consider a range of questions. Why does the population fluctuate from year to year? What determines the level of abundance? How strong is density dependence and at what life stage does it operate? What are the consequences of competitors, herbivores or predators on the population?

There are, however, considerable problems in answering such questions from observational data. Density manipulation (for example by adding or removing seeds) is invaluable in tackling such questions.

Box 1.6. **Population dynamics of the Brent Goose**

The Dark-bellied Brent Goose *Branta bernicla bernicla* breeds in arctic Russia and access to the breeding grounds is very difficult. Much has been learnt of the population biology simply by counting the birds every winter since 1955 in their European breeding grounds. Juveniles have pale bars on their wings and thus can also be counted separately.

The percentage of juveniles varies markedly between years from over half the population to less than 0.1%. This was assumed to be related to the variation in the snow cover in the breeding grounds, but Summers and Underhill (1987) showed that the percentage of juveniles follows a clear three-year cycle in which there is a good breeding success, followed by a poor breeding success, followed by an unpredictable season. It was suggested that this was linked to the lemming cycle in the Arctic with Arctic Foxes *Alopex lagopus* feeding on eggs and young when the lemming population was low. Subsequent field studies in Russia (Underhill *et al.* 1993) confirmed this hypothesis. The population increases markedly after a good breeding season and as a result of this research it has proved possible to predict the population changes in subsequent years.

References

Allport, G., Ausden, M., Hayman, P. V., Robertson, P. & Wood, P. (1989). The conservation of the birds of the Gola Forest, Sierra Leone. *ICBP Study Report No.* 39, ICBP, Cambridge.

Ausden, M. (1993). Ragwort control. *British Wildlife* 4, 378.

Bibby, C. J. & Hill, D. A. (1987) Status of the Fuerteventura Stonechat *Saxicola dacotiae*. *Ibis* **129**, 491–98.

Dolman, P. M. & Sutherland, W. J. (1991). Historical clues to conservation. *New Scientist* **1751**, 40–43.

Dolman, P. M. & Sutherland, W. J. (1992). The ecological changes of Breckland grass heaths and the consequences of management. *Journal of Applied Ecology* **29**, 402–13.

Howe, H. F. (1994). Response of early- and late-flowering plants to fire season in experimental prairies. *Ecological Applications* **4**, 121–33.

Summers, R. W. & Underhill, L. G. (1987). Factors related to breeding success of Brent Geese *Branta b. bernicla* and waders (Charadrii) on the Taimyr Peninsula. *Bird Study* **34**, 161–71.

Thomas, J. A., Thomas, C. D., Simcox, D. J. & Clarke, R. T. (1986). Ecology and declining status of the Silver-spotted Skipper butterfly (*Hesperia comma*) in Britain. *Journal of Applied Ecology* **23**, 365–80.

Underhill, L. G., Prys-Jones, R. P., Syroechkovski, E. E. jr., Groen, N. M., Karpov, V., Lappo, H. G., Van Roomen, M. W. J., Schekkerman, A., Spiekman, H. & Summers, R. W. (1993). Breeding of waders (Charadrii) and Brent Geese *Branta bernicla bernicla* at Pronchishcheva Lake, northeast Taimyr, Russia, in a peak and decreasing lemming year. *Ibis* **13**, 277–92.

2 Basic techniques

Jeremy J. D. Greenwood

British Trust for Ornithology, Thetford, Norfolk IP24 2PU, United Kingdom

Table 2.1. *Contents of Chapter 2*

Introduction

Objectives

The first step in any programme of research or monitoring is to define one's objectives. That statement may seem so obvious as not be worth making, yet many investigators rush into their work without having properly considered what are their objectives. As a result, the programme they devise may be inappropriate. At best, this will make their work inefficient; usually, it will result in them being unable to answer the question that they really mean to ask, though able to answer a related question; at worst, they may answer the related question without realising that it is not really the question they should have been answering.

The definition of objectives is particularly important in studies of the abundance of animals or plants because it will determine whether one needs to make a full count of the individuals present in an area (or, at least, an estimate of that number) or whether an index of numbers is satisfactory. By an index, we mean a measurement that is related to the actual total number of animals or plants – such as the number of eggs of an insect species found on a sample of leaves of its host plant (as an index of numbers of adult insects) or the number of rabbit droppings in a sample area (as an index of the number of rabbits). Because accurate counts are extremely difficult to make, one may have to make do with an index even when a count would be preferable. Indeed, so long as an index is sufficient for one's purposes, a reliable index is preferable to an unreliable count.

The objectives will also determine the extent to which a population should be divided by sex, age or size in one's study: some work on the structure of communities may require only that total numbers of each species are known, while population dynamicists will often need to know the composition of the population by sex and age.

If one's objectives include surveillance (observing changes over time), this may influence one's choice of method. Surveillance is considered briefly at the end of this chapter.

Know your organism

Having defined your objectives, the appropriate methods to apply will be determined by the characteristics of the species you are studying. It is therefore important to know as much as one can about the species and its ecology before planning one's work in detail. Some such knowledge can be obtained by reading the literature and by talking to experts but there is no substitute for carrying out preliminary observations oneself, preferably of the very population that one intends to study. The wisest ecologists tell their students to go out into the field in order to sit and watch their animals before they decide how to study them; wise students accept this advice.

Censuses and samples

It is sometimes possible to count the animals or plants in the whole of the population of interest (all the deer in a woodland, for example). Such a complete count is a true census. More often, one can study only part of the population, through taking representative samples (e.g. 10×10-m quadrats within a large forest, in order to study numbers of tree seedlings). A large part of this chapter is taken up with discussing how to arrange sampling so that the results are as accurate as possible.

Know the reliability of your estimates

It is sometimes possible to count a population directly. More often, one can only make an estimate of its true size. That being so, one wishes to know how accurate the estimate is; that is, how close it is to the true value. Accuracy is not to be confused with precision, which is the closeness of repeated estimates to each other. Estimates may be very close to each other (precise) but not at all close to the true value (because they are biased).

Measuring precision

For most of the methods of estimation given below, we also show how to calculate 95% confidence limits. A definition of these limits, to which a purist would object but which aids understanding, is that if the 95% confidence limits of an estimate of population size are CL_1 and CL_2, then the chances are 95% that the true population size lies between CL_1 and CL_2. It is important to provide confidence limits whenever an estimate is quoted, otherwise the reader has no idea how precise the estimate is. For comparison between different situations, it is often convenient to use the percentage relative precision (PRP: Box 2.1).

Box 2.1. **Percentage Relative Precision (PRP)**

Basics

$$\hat{N} = \text{population estimate}$$
$$CL_1, \ CL_2 = 95\% \text{ confidence limits of } \hat{N}$$

Definition

PRP is the difference between the estimated population size and its 95% confidence limits, expressed as a percentage of the estimate. Because confidence limits are sometimes asymmetrically distributed around the estimate, the mean difference between them and the estimate is used. This is calculated simply as

$$PRP = 50 \times (CL_2 - CL_1)/\hat{N}$$

Note

PRPs are only approximate guides. The simple methods presented in this chapter for calculating them are also often approximate, especially when numbers are small.

Sources of bias

Bias can arise through poor practical techniques. Thus even a direct count of a population is biased if the observer fails to see all the individuals. Other examples of poor practical technique are studying animals or plants only where they are common, applying marks (in order to be able to identify an animal subsequently) without considering whether marks are being lost, and causing increased mortality through using insensitive trapping techniques. In order that other workers (and you yourself, at a later date) can assess the probability that your working methods were a source of bias, you should always record them as fully and as carefully as possible.

Bias also arises because estimates are based on idealised statistical models, which do not properly reflect real life. Mark-and-recapture methods, for example, generally assume that all individuals in the population have equal chances of being caught; distance-measuring methods that the spatial distribution of individuals is random. These assumptions are often incorrect, so the estimates obtained are biased. In choosing a method, one must consider carefully whether its asumptions are likely to hold in one's study population and, if they are not, to what extent the results are likely to be biased as a result. Wherever possible, assumptions should be tested and biased estimates avoided. Do not be misled into believing that the more complex methods (which, necessarily, take up more space in this chapter) are generally better than simple ones. Since they rely on more restrictive assumptions, they are often worse. All other things being equal, simple methods are always to be preferred.

What if your estimate is biased?

The best solution is to find another method, or a modification of the existing method, that provides an apparently unbiased estimate. This may not be possible. Alternatively, it might be appropriate to use an index of population size rather than population number itself: the biased population estimate might actually be an appropriate index. At worst, one will be reduced to making the decision as to whether to reject the estimate entirely (and thus conclude that one has no idea as to the size of the population) or to accept it as a rough but biased guide. That decision will depend on one's judgement as to the likely magnitude of the bias, the value of having at least some idea of the size of the population (rather than none at all), and the cost of any wrong decisions that could be made as a result of basing conclusions on a false estimate of population size.

The balance of precision and cost

More precise estimates can always be obtained by extending the study but this will inevitably require additional resources, even if only in terms of the length of time devoted to the work. In planning any project one should consider either the PRP achievable for a given cost or, alternatively, the cost of achieving the level of PRP that one requires. It may turn out that the PRP achievable for the planned level of resources is insufficient to make the study worthwhile. In this case, one can decide either to spend more, in order to achieve the desired

PRP, or to abandon the study. In either case, by considering the PRP and costs ahead of the study, one will reduce the likelihood of wasting resources on a study that fails to deliver results that are precise enough.

Performing the calculations

All the calculations presented in this chapter can be performed on a simple pocket calculator. Some of them demand that the intermediate steps in the calculations are performed to a high degree of precision, so it is always best to use the full range of the calculator (though I have generally rounded numbers in the examples shown here, for ease of presentation). Note that in several cases there are more precise methods of calculation available than are shown here but these require powerful computer packages that are beyond the scope of this book.

Further reading

Seber (1982) is a key account of methods of estimating abundances, especially mark–recapture methods, though it is mathematically rather heavy.

Direct counts

Not as easy as it seems

It is sometimes possible to count directly the number of individuals in a study population, such as the number of acacia trees in a savannah area, the number of limpets on a section of rocky shore, or the number of elephants in a herd. But beware! Individual organisms (even trees and elephants) can turn out to be remarkably easy to overlook and many such direct counts turn out, when more careful studies are made, to be grossly incomplete. If one is familiar with a study area and frequently encounters the animals in it (especially if they become recognisable as individuals), one has a better chance of finding them all in a census; but such intimate knowledge should not lead one to be complacent about the possibility of missing animals that always take avoiding action as one enters their domain or that live in rarely visited parts of the study area.

It is important to avoid counting when weather conditions reduce the detectability of animals. Counting may be easier if animals are counted when concentrated together – such as migratory fish as they pass through a restricted part of a river, flocking birds on their roosts, or seals on their breeding beaches. The very crowding of the animals may, however, give rise to difficulties of counting under such conditions. One tip is to divide the area to be counted into sections, either using natural features or a grid that has been specially laid out. Photographs in which the animals can be counted at leisure are also useful. Plants and relatively immobile animals may be marked to indicate that they have already been counted.

Sampling the habitat

Small animals may be counted by removing a sample of the habitat (such as a volume of water, a core of soil, or some leaves of a plant) and sorting it later in the laboratory, in comfortable conditions, with good light and perhaps using optical aids. It is important not to let animals escape when one collects such samples and to store the samples so that, even if the animals die, they remain identifiable. Extracting the animals from such samples may be time-consuming, so various mechanical aids such as Tullgren funnels (see p. 160) are often used. It is important always to check, for the particular animals and substrate involved in one's study, that all animals are being extracted – by checking what remains in some of the samples after they have passed through the extractor. (An extractor that is less than 100% efficient may, of course, be used to provide an index of the animal population so long as it extracts a constant proportion of the animals from the sample.)

When animals are unexpectedly numerous, a mechanical extractor may produce such large numbers that it is not cost-effective to count them all. Subsamples may then be taken, to reduce the labour. It is neither necessary nor desirable that these should be random samples: what is important is that they should be representative. Because animals floating in a dish, for example, tend to cluster together (often in the corners or towards the centre), it needs care to achieve representative subsampling. It is usually necessary to mark out a grid and to take subsamples from all parts of it. Southwood (1978, p. 145) shows a useful grid for subsampling one-sixth of the total from a circular dish.

Complete enumeration over time

There are some species in which it is possible to recognise individuals. This means that, even if one never sees all the members of a population at once, one can count the number in the population simply by keeping a record of all the different individuals seen. Eventually, unless there is a constant stream of immigrants or of new-born animals, one will reach a stage when no new animals are being seen; the list of the individuals seen is then a complete list of the population. More commonly, one can mark animals when they are seen for the first time; by counting the number of animals marked and carrying on until one is encountering no unmarked animals, the total number present can again be determined. (Marking methods are described for various sorts of animals in subsequent chapters.)

An alternative way of counting the population over time is to catch the animals and not release them – as might be done in a pest-control programme. Again, if the trapping is continued until no further animals are being caught, the total catch tells one what the population size was at the start of the study. Immigration is a particular problem with this method, since new animals are likely to move into the living space vacated by the trapped animals.

The problem with enumeration is the effort that is required if one is to be confident that one has observed every animal in the population. For example, if one has encountered only 10 different individuals, one has to carry on the observations until there have been a total of 50 separate encounters (an average of 5 per individual) before one can be 90% confident that there are, indeed, only 10 animals present in the population. The methods of the next section, which allow one to estimate the number of unobserved animals, are more efficient.

Mark–recapture methods

Fundamentals of mark–recapture

The basic idea

Suppose that one catches a sample of animals from a population, marks them and releases them. Suppose further that, after allowing the marked animals time to become thoroughly mixed into the rest of the population, one takes another sample. It is reasonable to assume that the proportion of marked animals in the second sample is the same as that in the population at large. This idea can be expressed symbolically as follows. If n_1 is the number of animals first marked and released, if n_2 is the size of the second sample and m_2 the number of marked animals in that sample, and if N is the total population size, then we expect: $m_2/n_2 = n_1/N$. It is obvious that, since n_1, n_2, and m_2 are known, N can be estimated. All mark–recapture methods rest on this basic idea, though most entail animals being caught and marked on several occasions.

When marking or trapping are unnecessary

Some animals can be recognised individually through natural differences in appearance: marking is then unnecessary. Sometimes they can be recognised without being caught. In this case, one can simply observe a sample of n_1 animals at one time, listing which ones are present; then one observes another sample (n_2) and identifies which animals were present in both samples (m_2), in order to estimate the total population size as above.

Since both trapping and marking may affect the animal's behaviour (and even its survival), which can lead to bias in estimating population size, methods that depend on natural distinguishing features and that avoid the need for trapping are to be preferred. Sometimes, as a compromise, one is able to mark animals in such a way that the marks can be recognised without recapturing the animals – though there is the danger that marks may then be overlooked or that they will be so conspicuous as to increase the probability of the animals being taken by predators.

Marking

Methods of marking different sorts of animals are described in subsequent chapters. Some methods, such as the numbered rings applied to birds, allow each individual to be identified whenever it is caught. Other methods, such as applying a red dot of paint to the underwing of moths caught in the first sample, a blue dot to those caught in the second sample, a green dot in the third, etc. (so that an animal caught on all three occasions would carry three dots, of different colours), result only in 'batch-specific' marking. Both individual and batch-specific marks, since they allow each animal's capture history to be determined, are suitable for all mark–recapture methods, though individual marking can provide valuable supplementary information.

It is sometimes impossible to apply individual or batch-specific marks: animals are then identifiable only as having been caught at least once before (marked) or as not previously having been caught (unmarked). This limits the range of methods that can be used to estimate population size (see below).

All methods assume that marks are not lost. This can be checked by marking animals in two different ways. If any of them have only one mark when they are subsequently recaptured, this shows that the other one has been lost. If one assumes that the chances of losing the two types of mark are independent, then the proportion of type A marks lost between marking and recapture can be estimated from the numbers of animals recaptured with both marks (R_{AB}) and with only type B (R_B). It is: $R_B/(R_B + R_{AB})$. Such estimates of the rates of loss of marks can be used to correct population estimates made by some of the methods described below. If marks are lost and no correction is applied, population size will be overestimated. The same bias will arise if marks are overlooked when animals are recaptured. This is a problem in fisheries biology, for example, where the scientist may mark a sample of fish but then rely on commercial fishing for the recapture information.

Marking is also assumed not to affect mortality or behaviour. This assumption must be checked carefully because population estimates are seriously biased if marks do affect the animals. The mortality rates of animals marked in different ways can be compared to determine if any of the methods are particularly harmful, while direct observations can often be made of the behaviour of marked and unmarked animals. Statistical tests can also be applied to the recapture data themselves.

Variation in catchability

Whether an animal is caught or not depends on a variety of circumstances. Because variation in the likelihood of capture can produce huge biases in the estimates of population size, it is important to recognise sources of that variation.

The proportion of the population captured can vary from time to time depending both on uncontrollable factors such as weather and season and on the trapping effort (number of traps, time for which they are set, etc.). Although such variation sometimes produces little or no bias (unless the probability of capture tends systematically to increase or decrease over the period of the study), it does reduce the precision of the estimates. It is generally best, therefore, to keep trapping effort constant over all capture occasions. The Burnham & Overton method (see below) is seriously biased by any temporal variation in capture rate.

All methods assume that all the animals in the population can be trapped. If some of the animals are too small to be caught in the traps, for example, one's estimate of population size will refer only to those animals that are large enough to be trapped. Similarly, animals that are aestivating or hibernating during the period over which the population is studied will not be included in the population estimate. More generally, the methods assume that all animals in the population are equally trappable. If they are not, population size will be seriously underestimated. Differences in trappability are most likely, but fortunately most readily detectable, when there are identifiable subgroups in the population, such as different sexes or ages. In this case, it is generally better to estimate the numbers in each subgroup separately rather than to pool the data to obtain an overall estimate. Only if the differences in trappability between subgroups are slight and sample sizes are small is it better to pool all the data and make a direct estimate of total population size. Subgroups should not be pooled without formally testing the differences between them but we have not included such tests in the account below because pooling is so rarely to be recommended.

It is important that differences in catchability are not created by the investigator. If too few traps are used, some animals' home ranges may not include a trap. If insufficient time is allowed between marking and recapture, the marked animals will not mix completely with the rest of the population. Both of these problems can be reduced by having a high density of traps and by randomising their positions. The last can be achieved by adapting the methods used for random sampling described later (p. 75).

Trap responses are also important. Once caught, animals may become trap-shy, reluctant to enter a trap again, avoiding its location, or even leaving the area completely. Sometimes animals become trap-happy, especially if traps are baited. Both types of response may be short-lived or permanent. Trap-shyness results in overestimates of population size, trap-happiness in underestimates. The chances of trap-shyness developing can be reduced by employing methods that are not likely to distress the animals. If baiting is essential – either to attract the animals to the traps or to keep them alive after they have been caught – it is sometimes possible to set the traps open and baited (but fixed so that no animals entering them will be caught) for some time before the first catch is made. This allows all animals in the population to become equally trap-happy before catching starts.

The impact of heterogeneity of the probability of capture is so great that it is essential to carry out statistical tests of the homogeneity assumption whenever possible. The tests appropriate to each method of estimation are presented below. Should these tests prove significant heterogeneity to exist, and should no alternative method of population estimation be available, one must accept that any estimate made is probably severely biased. It may, depending on circumstances, be useful to take it as a rough guide to the true population size or possible to use it as an index of population size; what one must not do is to take it as an accurate estimate.

Open or closed populations

Many of the methods described in this section assume the population to be 'closed'. That is, they assume that there are no gains (births or immigration) or losses (deaths or emigration) during the course of the study. If these methods are applied to 'open' populations (subject to gains or losses), the estimates of population size will generally be biased (usually upwards), though some (which will be identified below) are not affected by losses. If one uses one of the methods appropriate for a closed population, it is important to minimise the chances of losing or gaining individuals by conducting the study over a short period and at a time of year

Table 2.2. *Mark–recapture methods presented in this book*

	Petersen	Schnabel	Burnham & Overton	Removal	Jolly–Seber
Number of capture/recapture sessions	Two	Several	Several	Several	Several
Population open or closed	Closed	Closed	Closed	Closed	Open
Capture probabilities	Uniform	Uniform	Heterogeneous	Trap response	Uniform
Type of mark required	Single	Single	Multiple or individual	Single	Batch-specific or individual

when births, deaths, and movements are likely to be few. It may be possible to recognise new-born individuals and exclude them from the calculations. Immigration and emigration are more difficult to deal with, though it is sometimes possible to measure the rates and allow for them.

If one applies closed population models to open populations, the estimate of population size is biased. If one applies open models to closed populations, the estimate is not biased. Why do we not therefore apply open models routinely, to both open and closed populations, and thus avoid the possibility of obtaining a biased estimate by applying a closed model to a population that we do not realise is actually open? The answer is that, because open population models make fewer assumptions (that is, they do not assume rates of gain and loss to be zero), the estimates of population size that they produce are less precise. Nonetheless, if one cannot be sure that a population is closed it is better to use an open model: an imprecise but unbiased estimate is generally preferable to a precise but biased estimate.

Increasing precision

This makes it especially important to estimate the PRP (Box 2.1) likely to be achieved in advance of one's study. It is important to recognise that mark–recapture methods are often disappointingly imprecise, given the amount of effort involved. Precision is increased if a high proportion of the population is caught. Catching a very high proportion has the further benefit of limiting any bias that might arise through assumptions being violated.

Overview of mark–recapture methods

Various mark–recapture methods are available, each involving different assumptions or different statistical methods for arriving at an estimate of population size. Those considered here (summarised in Table 2.2) include all of the more widely used methods. Each is suitable for a particular set of circumstances, corresponding to its particular assumptions. Your choice of method will depend on which assumptions are likely best to fit your study population, and on practical considerations such as the number of trapping sessions and whether it is possible to apply multiple or individual marks.

References

Seber (1982) and Pollock *et al.* (1990) consider mark–recapture methods in detail. I have presented further references under particular methods below, as necessary.

The Petersen method

This, also known as the Lincoln index, is the most basic method. It involves one session of catching and marking and one recapture session. If there are losses from the population, the estimate obtained is for the size of the population at the time of the first catch; if there are gains, the estimate corresponds to population size during the second catch; if there is turnover (gains and losses) the estimate is biased. Heterogeneity between animals in catchability and trap responses also causes the usual biases. Differences in the proportion of the population caught on the two occasions have no effect on the estimate (though it is generally most cost-effective to make the same catching effort on both occasions). It is not possible to use the mark–recapture data themselves to check the assumptions of this method, so one must exercise particular caution when using it.

The method of estimating population size, with approximate confidence limits, is shown in Box 2.2. Note that, to eliminate bias arising for statistical reasons, the estimate of population size is not simply the common-sense value, $n_1 n_2 / m_2$.

Box 2.2. **The Petersen method**

Basics

> n_1 = number marked and released on the first occasion
> n_2 = total number caught on the second occasion
> m_2 = number of marked animals found on the second occasion

The total population size (N) is estimated as

$$\hat{N} = (n_1 + 1)(n_2 + 1)/(m_2 + 1) - 1$$

Planning

For this method, the percentage relative precision (Box 2.1) is given by

$$Q = 200\sqrt{N/n_1 n_2}$$

and

$$n = 200\sqrt{N}/Q \quad \text{(if } n_1 = n_2 = n)$$

Thus, the PRP attainable if one believes that the population size is around 40 000 and that one is able to take samples of roughly 1500 is

$$Q = 200\sqrt{40\,000/1500^2} = 27\%$$

If one wishes to obtain a PRP of 10%, then the required sample sizes are

$$n = 200\sqrt{40\,000}/10 = 4000 \quad \text{(for each sample)}$$

Example

Climbing cutworms (larvae of various species of noctuid moths) in a field of blueberries *Vaccinium* in New Brunswick, Canada (Wood 1963).

$$n_1 = 1000, \ n_2 = 1755, \ m_2 = 41$$

Estimate of total population size

$$\hat{N} = [(1001)(1756)/42] - 1 = 41\,851 - 1 = 41\,850$$

Approximate confidence limits of the estimate

Calculate

$$p = m_2/n_2$$
$$= 0.023$$

Calculate the two values

$$W_1, \ W_2 = p \pm \left[1.96 \sqrt{p(1-p)\left(1 - \frac{m_2}{n_1}\right)\Big/\left(n_2 - 1\right)} + \frac{1}{2n_2} \right]$$
$$= 0.023 \pm [1.96\sqrt{(0.023)(0.977)(0.959)/1754} + 1/3510]$$
$$= 0.023 \pm 0.007$$
$$= 0.030 \text{ and } 0.016$$

Divide W_1 and W_2 into n_1 to obtain approximate 95% confidence limits for $\hat{N}$.

In the cutworm example these are

$$1000/0.030 = 33\,000$$

and

$$1000/0.016 = 63\,000$$

If the number of marked animals found in the second sample (m_2) is less than 8 then the estimate of N is biased. The confidence limits calculated according to Box 2.2 are reasonably accurate for $m_2 > 50$. Methods are available for calculating more accurate confidence limits for smaller values of m_2 but are perhaps only worth using if the proportion of recaptures (p) is high (when the confidence range will be relatively narrow) and when one is reasonably sure that the assumptions of the methods are met in one's example (since it is only worth being much concerned about precision if there is no bias).

The Schnabel method

The Schnabel method rests on the same assumptions as the Petersen but is appropriate to situations in which animals are captured on several occasions; all unmarked animals in each capture are marked before being released. The method depends simply on observing how the proportion of marked animals in catches increases as more animals have been marked; when this proportion equals 1.0, the total number of animals previously marked must be the number in the population (Figure 2.1). Box 2.3 shows how to calculate an estimate of the population size, with confidence limits. Note that, for a given total effort, more precise estimates are obtained if the effort is the same throughout the study than if it varies.

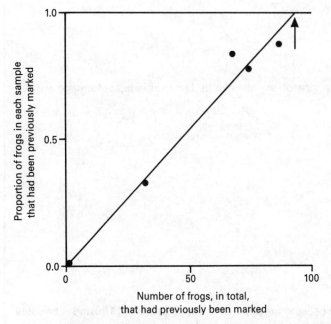

Figure 2.1 The Schnabel method applied to a population of Cricket Frogs *Acris gryllus* in Louisiana, USA, sampled over five successive days. (Data from Turner 1960.) The arrow indicates the estimated population size.

Box 2.3. **The Schnabel method**

Data may be analysed in various ways. The regression method of Schumacher and Eschmeyer is presented here.

Basics

$$S = \text{number of samples}$$
$$n_i = \text{number of animals in the } i\text{th sample}$$
$$m_i = \text{numer of animals in the } i\text{th sample that are already carrying marks}$$
$$u_i = n_i - m_i = \text{number of unmarked animals in the } i\text{th sample}$$
$$M_i = \sum_{j=1}^{i-1} u_j = \text{number of animals marked prior to the } i\text{th sample}$$

Preliminary calculations:

$$A = \sum n_i M_i^2$$
$$B = \sum m_i M_i$$
$$C = \sum m_i^2 / n_i$$

Total population size is estimated as

$$\hat{N} = A/B$$

Planning

For this method, provided the size of the samples is fairly uniform, percentage relative precision (Box 2.1) is given by

$$Q = 200\sqrt{N/(S\bar{n}^2)}$$

and

$$\bar{n} = 200\sqrt{N/S}/Q$$

and

$$S = N[200/\bar{n}Q]^2$$

where $\bar{n}$ is the average sample size.

Thus, the PRP attainable if one believes the population size to be around 100 and achievable sample sizes about 50 (and if one plans to take 5 samples) is

$$Q = 200\sqrt{100/(5 \times 50^2)} = 18\%$$

If one wished to obtain a PRP of 10% with the same number of samples, then their average size would need to be

$$n = 200\sqrt{100/5}/10 = 89$$

If one wished to obtain a PRP of 10% with sample sizes of about 50, then the number of samples required would be

$$s = 100[200/(50 \times 10)]^2 = 16$$

Example

The Cricket Frog data presented in Figure 2.1:

i	n_i	m_i	u_i	M_i	$n_i M_i^2$	$m_i M_i$	m_i^2/n_i
1	32	0	32	0	0	0	0
2	54	18	36	32	55 296	576	6.0000
3	37	31	6	68	171 088	2108	25.9730
4	60	47	13	74	328 560	3478	36.8167
5	41	36	5	87	310 329	3132	31.6098
					$A = 865\,273$	$B = 9294$	$C = 100.3995$

Estimate of total population size

$$\hat{N} = 865\,273/9294 = 93.10 = 93$$

Confidence limits of the estimate

95% confidence limits are given by

$$A\left/\left[B \pm t\sqrt{(AC - B^2)/(S-2)}\right]\right.$$

where t is Student's t for $S-2$ degrees of freedom at the 5% significance level. Confidence limits for the Cricket Frog estimate are

$$865\,273/[9294 \pm 3.182\sqrt{(865\,273 \times 100.3995 - 9294^2)/3}]$$
$$= 865\,273/[9294 \pm 1291.93]$$
$$= 82 \text{ and } 108$$

A test of goodness-of-fit to the model

The expected values of m_i and u_i are calculated from

$$\text{Exp}(m_i) = M_i n_i/\hat{N}$$
$$\text{Exp}(u_i) = n_i - \text{Exp}(m_i)$$

Note that one should use the precise estimate of $\hat{N}$, rather than the value rounded to a whole number.

The observed and expected values should be tabulated alongside each other, leaving out the first sample. In the current example

Sample		2	3	4	5
Observed values	m_i	18	31	47	36
	u_i	36	6	13	5
Expected values	m_i	18.56	27.02	47.69	38.31
	u_i	35.44	9.98	12.31	2.69

Note that in this example one of the expected values is rather small; however, the test usually performs satisfactorily so long as no more than 20% of the expected values are less than 5 and none is less than 1. If there are too many small values, adjacent columns of the table of observed and expected values should be added together.

Next calculate

$$G = 2\sum O_j \log_e (O_j/E_j)$$

where

O_j = the jth observed value (including both m_i and u_i values)
E_j = the jth expected value (including both m_i and u_i values)

In this case

$$G = 2[18\log_e(18/18.56) + \ldots 5\log_e(5/2.69)] = 4.20$$

G should be compared with the tabulated value of χ^2, with $S-2$ degrees of freedom (3 in this case). The value of 4.20 is not significant, indicating that the regression line is a satisfactory fit to the data.

In making a Schnabel estimate, it is important to plot a graph such as Figure 2.1 because if the points do not fall on a straight line then the estimate of population size is likely to be biased. The appropriate line is that drawn from the origin to the point at which $M_i = \hat{N}$ and the proportion of animals marked = 1.0. A formal test of the model assumptions may be made by assessing the goodness-of-fit of the data to the regression line, as shown in Box 2.3. The basic Schnabel method and this test require only that animals are marked (to indicate that they have been caught at least once); if a more complex marking system that enables one to identify how many times each animal has been caught is used (such as batch-specific or individual marks) then a generally more sensitive test is the goodness-of-fit to a zero-truncated Poisson distribution, which is the distribution that would describe the number of animals caught 1, 2, 3 … times if all the assumptions of the model were true (Box 2.4).

Box 2.4. **Test of goodness-of-fit to a zero-truncated Poisson distribution**

Basics

One must first estimate the mean of the underlying Poisson distribution ($\bar{x}$) from the observed mean number of times that each animal has been caught ($\bar{x}_T$). For observed means less than 2.0, use Table 2.3; for larger values estimate $\bar{x}$ as

$$\bar{x}_T - Z - Z^2 - 1.5Z^3 - 2.6Z^4 - 5.2Z^5$$

where

$$Z = \bar{x}_T e^{-\bar{x}_T}.$$

As an example of this calculation, suppose that $\bar{x}_T = 2.540$. Then

$$Z = 2.540e^{-2.540} = 0.200$$
$$\bar{x} = 2.540 - 0.2 - (0.2)^2 - 1.5(0.2)^3 - 2.6(0.2)^4 - 5.2(0.2)^5 = 2.282$$

Having estimated $\bar{x}$ from the table or by calculation, the expected number of animals caught exactly i times is calculated from the equation for the Poisson distribution; the observed and expected numbers are used to calculate a G value in the usual way.

Example

Agamid lizards *Amphibolurus barbatus* studied over 21 sampling occasions in Australia by J.A. Badham (Caughley 1977): f_k is the number of animals caught exactly k times.

k	1	2	3	4	5	6	7	8	9	10–21	Sum
f_k	23	7	3	2	1	0	1	0	1	0	38
$f_k k$	23	14	9	8	5	0	7	0	9	0	75

The observed mean, $\bar{x}_T = \sum f_k k / \sum f_k = 75/38 = 1.97$

The corresponding Poisson mean ($\bar{x}$) from Table 2.3 is 1.553

The expected number of lizards caught exactly once is

$$E_1 = \left[\frac{(\sum f_k)e^{-\bar{x}}}{1 - e^{-\bar{x}}} \right] \bar{x} = \left[\frac{(38)e^{-1.553}}{1 - e^{-1.553}} \right] 1.553 = 15.840$$

Successive values are calculated from

$$E_2 = E_1 \bar{x}/2 = 15.840 \times 1.553/2 = 12.300$$
$$E_3 = E_2 \bar{x}/3 = 12.300 \times 1.553/3 = 6.367$$
etc.

One calculates successive values of E_i until the sum of all the values already calculated is within 5 of the total number of animals captured $(\sum f_k)$. In this example, $E_1 + E_2 + E_3 = 34.507$. The expected number of animals caught more often than this is obtained by subtraction:

$$E_{4 \text{ or more}} = 38 - 34.507 = 3.493$$

The values are tabulated, combining adjacent values so that none of the expected values is less than 5:

k	1	2	3 or more	Sum
$O_k = f_k$	23	7	8	38
E_k	15.84	12.30	9.86	38

Calculate G as usual:

$$G = 2\sum O_k log_e(O_k/E_k)$$

Test this against χ^2 with $c - 2$ degrees of freedom, where c is the number of columns in the table (after combining to avoid expected values less than 5).

In this case

$$G = 2[23log_e(23/15.84) + 7log_e(7/12.30) + 8log_e(8/9.86)]$$
$$= 5.92, \text{ with 1 degree of freedom}$$

This is significant at the 5% level, indicating that the frequency of captures does not fit a zero-truncated Poisson distribution.

Reference
Caughley (1977).

If either test is significant, indicating that the model assumptions are violated, one needs first to consider what the violations might be. If the points are simply rather widely scattered about the line rather than falling on a curve (which suggests variation in catchability between trapping sessions but neither a trend nor heterogeneity in catchability), then the Schnabel estimate will be satisfactory. If the points tend to fall on a curve, it will not. In this case, if one suspects that the population is open rather than closed, then one should switch to the Jolly–Seber method, but if one is convinced that the population is closed, then the Burnham & Overton or removal methods should be used, as appropriate.

The Burnham & Overton method

This method allows animals to differ in probability of capture; the assumption that they do not is perhaps the commonest cause of bias in capture–recapture studies that use other methods. The information used to estimate population size in this case is the number of

Table 2.3. *Relationship between mean of a Poisson distribution ($\bar{x}$, in the body of the table) and the mean of the equivalent zero-truncated Poisson distribution ($\bar{x}_T$), for values of $\bar{x}_T = 1.01$ to 1.99. (For larger values, use the method of calculation shown in Box 2.3.)*

To find $\bar{x}$ corresponding to $\bar{x}_T = 1.23$ (for example), enter the table on the row marked 1.2 and the column marked 0.03 (which gives $\bar{x} = 0.430$).

$\bar{x}_T$	0.00	0.01	0.02	0.03	0.04	0.05	0.06	0.07	0.08	0.09
1.0	—	0.020	0.040	0.060	0.079	0.099	0.118	0.137	0.156	0.175
1.1	0.194	0.213	0.232	0.250	0.268	0.287	0.305	0.323	0.341	0.359
1.2	0.377	0.395	0.412	0.430	0.447	0.465	0.482	0.499	0.516	0.533
1.3	0.550	0.567	0.584	0.601	0.617	0.634	0.650	0.667	0.683	0.700
1.4	0.715	0.732	0.748	0.764	0.780	0.795	0.812	0.828	0.843	0.859
1.5	0.875	0.890	0.906	0.921	0.936	0.952	0.967	0.982	0.997	1.012
1.6	1.027	1.042	1.057	1.072	1.087	1.102	1.117	1.131	1.146	1.161
1.7	1.175	1.190	1.204	1.219	1.233	1.247	1.262	1.276	1.290	1.304
1.8	1.318	1.333	1.347	1.361	1.375	1.389	1.403	1.417	1.430	1.444
1.9	1.458	1.472	1.485	1.499	1.513	1.526	1.540	1.553	1.567	1.580

animals caught exactly 1, 2, 3 and 4 times over the entire study, which should comprise more than four sample periods. The critical assumption in using these data is that capture probabilities do not change over time, so it is important to maintain constant trapping effort and to adopt procedures that minimise the likelihood of trap responses. Even when great care is taken, the assumption of constant capture probabilities should always be tested.

Box 2.5 presents the calculations, which involve making four different estimates of population size and then testing successive values to choose the best (defined as the one which minimises the mean square error – that is, the variance plus the square of the bias).

Box 2.5. **The Burnham & Overton method**

Basics

S = number of samples
f_k = number of animals caught exactly k times

Example

Cottontail Rabbits *Sylvilagus floridanus* in an experimental enclosure in USA (Edwards & Eberhardt, 1967), sampled on 18 occasions.

k	1	2	3	4	5	6	7	8–18	Sum ($\sum f_k$)
f_k	43	16	8	6	0	2	1	0	76

Estimate of total population

Calculate α coefficients from

$\alpha_{11} = (S-1)/S$

$\alpha_{12} = (2S-3)/S \qquad \alpha_{22} = \qquad -(S-2)^2/S(S-1)$

$\alpha_{13} = (3S-6)/S \qquad \alpha_{23} = -(3S^2-15S+19)/S(S-1) \qquad \alpha_{33} = \qquad (S-3)^3/S(S-1)(S-2)$

$\alpha_{14} = (4S-10)/S \qquad \alpha_{24} = -(6S^2-36S+55)/S(S-1) \qquad \alpha_{34} = (4S^3-42^2+148S-175)/S(S-1)(S-2)$

$$\alpha_{44} = -(S-4)^4/S(S-1)(S-2)(S-3)$$

In this example

$\alpha_{11} = 0.9444$

$\alpha_{12} = 1.8333 \qquad \alpha_{22} = -0.8366$

$\alpha_{13} = 2.6667 \qquad \alpha_{23} = -2.3562 \qquad \alpha_{33} = 0.6893$

$\alpha_{14} = 3.4444 \qquad \alpha_{24} = -4.4150 \qquad \alpha_{34} = 2.4937 \qquad \alpha_{44} = -0.5231$

Calculate β values as the difference between each α value and the one above it in the table ($\beta_{ij} = \alpha_{ij} - \alpha_{ij-1}$). Note that β_{11} is not required, although all other values of β_{ij} for which $i=j$ are required. In this example

$\beta_{12} = 0.8889 \qquad \beta_{22} = -0.8366$

$\beta_{13} = 0.8333 \qquad \beta_{23} = -1.5196 \qquad \beta_{33} = 0.6893$

$\beta_{14} = 0.7778 \qquad \beta_{24} = -2.0588 \qquad \beta_{34} = 1.8044 \qquad \beta_{44} = -0.5231$

Calculate four estimates of N ($\hat{N}_j$, $j = 1, 2, 3, 4$), from

$$\hat{N}_j = \sum f_k + \sum \alpha_{ij} f_k$$

In this example

$\hat{N}_1 = 76 + 0.9444(43) \hspace{5.5cm} = 116.6$

$\hat{N}_2 = 76 + 1.8333(43) - 0.8366(16) \hspace{3.4cm} = 141.4$

$\hat{N}_3 = 76 + 2.6667(43) - 2.3562(16) + 0.6893(8) \hspace{1.4cm} = 158.5$

$\hat{N}_4 = 76 + 3.4444(43) - 4.4150(16) + 2.4937(8) - 0.5231(6) = 170.3$

Test the differences between successive estimates using

$$t_j = (\hat{N}_{j+1} - \hat{N}_j)\sqrt{(\sum f_k - 1)/\{\sum f_k(\beta_{j,j+1}^2 f_j) - (\hat{N}_{j+1} - \hat{N}_j)^2\}}$$

If $t_j > 1.96$, the higher estimate is significantly better than the lower. It should then be tested against the next higher estimate, continuing until one encounters a non-significant different or only $\hat{N}_4$ remains (in which case it is taken as the best estimate).

In this example

$$t_1 = (141.4 - 116.6)\sqrt{75/\{76[(0.8889)^2 43 + (0.8366)^2 16] - (141.4 - 116.6)^2\}}$$
$$= 4.05, \text{ which is significant}$$
$$t_2 = 2.02, \text{ which is significant}$$
$$t_3 = 1.07, \text{ which is not significant, so } \hat{N}_3 \text{ (rounded to the nearest whole number) can be taken}$$
as the best estimate of N in this case.

Approximate confidence limits of the estimate

Approximate 95% confidence limits of the estimate $\hat{N}_j$ are given by

$$\hat{N}_j \pm 1.96\sqrt{\sum [f_i(\alpha_{ij}+)^2] - \hat{N}_j}, \text{ summation being over all } i$$

In this case, these are

$$158.5 \pm 1.96\sqrt{43(3.667)^2 + 16(1.3562)^2 + 8(1.6893)^2 - 158.5}$$
$$= 158.5 \pm 41.7$$
$$= 117 \text{ and } 200$$

Testing the assumption that capture probability is constant over time

If this assumption is true, then one expects an equal number of animals be caught on each occasion. The observed and expected numbers can be compared in a standard goodness-of-fit test.

In this example, the numbers were

i	1	2	3	4	5	6	7	8	9	10	11	12	13	14	15	16	17	18	Sum
n_i	9	8	9	14	8	5	18	11	4	3	16	5	2	7	9	0	4	10	142

Hence, the expected numbers on each day were $142/18 = 7.8889$.

$$G = 2\sum O_i \log_e(O_i/E_i)$$
$$= 2[9\log_e(9/7.8889) + 8\log_e(8/7.8889) \dots 10\log_e(10/7.8889)]$$
$$= 55$$

This has $S-1$ degrees of freedom (17 in this case), so the result is highly significant, indicating that the mean catchability was not constant from day to day. As a result, the estimate of population size is likely to be biased.

References

Burnham & Overton (1978, 1979).

This method appears to need particularly high capture rates to achieve precise estimates: the wide confidence limits obtained in Box 2.5 are a consequence of only about 10% of the population being caught each day. Unfortunately, there is no general equation for calculating the PRP for this method.

The removal method

This is the only method that can safely be applied to closed populations in which there is a trap response. It can be used in such circumstances because it does not use information from recaptures. It is not, therefore, a mark–recapture method and will be dealt with in detail in the next section (p. 40).

The method was devised for studies in which trapped animals were removed from the population and is based on the rate of decline in numbers trapped as the population is reduced by the removal of animals. If animals are marked rather than being removed, the method uses the decline in number of unmarked animals to estimate population size.

Variation in trapping effort can be allowed for in the calculations but estimates are biased if catchability varies between animals.

The Jolly–Seber method

The general method

This is the method of choice for open populations. It requires at least three samples to be taken and marks that are at least batch-specific, since one needs to know not just how many animals in each sample have been previously marked but also the most recent previous sample in which each of them was last trapped. If animals are individually identifiable, this information can be derived from the capture histories of each – i.e. the list of samples in which it was caught. If marks are merely batch-specific, one must record the number of animals with each possible capture history. Box 2.6 presents an example: it shows that, in the fourth sample, the number of animals never previously caught was 40, the number caught before on the first (but not the second and third) occasion was 10, and so on.

Box 2.6. **An example of how to record the composition of each sample for the Jolly–Seber method when marks are batch-specific**

Plus signs indicate that an animal was caught on that day, zeros that it was not. This example refers to sample number 4 in a hypothetical study.

Days on which previously captured			Number of animals
1	2	3	
0	0	0	40
+	0	0	10
0	+	0	9
0	0	+	11
+	+	0	2
+	0	+	2
0	+	+	1
+	+	+	0

The method not only allows for gains and losses but provides estimates of the number of animals entering the population and the survival rate, though the former includes both immigrants and births and 'survival' is the complement of both emigration and mortality.

The method is presented in Box 2.7, with a method for testing the underlying assumptions. These are that there is no heterogeneity between animals in their catchability, that there is no trap response, that catching and marking do not affect mortality or emigration rates, and that emigration is permanent. As with all mark–recapture methods, heterogeneity between animals in catchability can produce severe negative biases in the estimates of population size, unless each sample includes a high proportion (i.e. over 50%) of the population. The goodness-of-fit test in Box 2.7 is a reduced version of the test proposed by Pollock *et al.* (1990). It depends on inspecting the data from each sample for evidence that animals first captured in that sample are more or less likely to be recaptured than animals that had been caught in previous samples (as well as the sample in question). Such associations would arise if animals differed in trappability or if there were certain sorts of trap response, both potentially serious sources of bias.

Since the values calculated using this method are estimates, it is possible for them to fall outside realistic ranges: Φ_i may be greater than 1.0 and $\hat{B}_i$ may be negative. In such cases, values of 1.0 and 0.0 should be taken as the best biologically realistic estimates. Jolly–Seber estimates are very imprecise unless the number of marked animals in each sample is more than 10. Unfortunately, there is no general equation for calculating the PRP for this method.

Box 2.7. **The Jolly–Seber method**

Definitions

n_i = total number of animals caught in the ith sample

R_i = number of animals that are released after the ith sample

m_i = number of animals in ith sample that carry marks from previous captures

m_{hi} = number of animals in the ith sample that were most recently caught in the hth sample

Example

Black-kneed Capsids *Blepharidopterus angulatus* caught at 3- or 4-day intervals in a British apple orchard (Jolly 1965).

The data are best summarised in a table of m_{hi} values, with n_i and R_i values across the top and the m_i values (which are the sums of the m_{hi} values in the column above) across the bottom. Thus in the table below, 169 insects were caught in sample 3 (n_3) of which 164 (R_3) were released; 3 of them bore marks from sample 1 (but not from sample 2) (m_{13}) and 34 bore marks from sample 2 (m_{23}), giving a total of 37 marked insects (m_3).

i	1	2	3	4	5	6	7	8	9	10	11	12	13	r_h	z_i
n_i	54	146	169	209	220	209	250	176	172	127	123	120	142		
R_i	54	143	164	202	214	207	243	175	169	126	120	120	—		
h															
1		10	3	5	2	2	1	0	0	0	1	0	0	24	—
2			34	18	8	4	6	4	2	0	2	1	1	80	14
3				33	13	8	5	0	4	1	3	3	0	70	57
4					30	20	10	3	2	2	1	1	2	71	71
5						43	34	14	11	3	0	1	3	109	89
6							56	19	12	5	4	2	3	101	121
7								46	28	17	8	7	2	108	110
8									51	22	12	4	10	99	132
9										34	16	11	9	70	121
10											30	16	12	58	107
11												26	18	44	88
12													35	35	60
m_i	0	10	37	56	53	77	112	86	110	84	77	72	95		

The table contains two other sets of summations:

r_h = the number of animals that were released from the hth sample and were subsequently recaptured; these are simply the row sums

z_i = the number of animals caught both before and after the ith sample but not in the ith sample itself; z_i is the sum of all the m_{hi} that fall in columns to the right of column i and all rows above row i; thus the dashed lines in the table delimit the m_{hi} values that must be summed to obtain z_4, for example

Parameter estimates

$\hat{M}_i$ = number of marked animals in the population when the ith sample is taken (but not including animals newly marked in the ith sample).

$= m_i + (R_i + 1)z_i/(r_i + 1)$

$\hat{N}_i$ = population size at the time of the ith sample

$= \hat{M}_i(n_i + 1)/(m_i + 1)$

Φ_i = proportion of the population surviving (and remaining in the study area) from the ith sampling occasion to the $(i+1)$th

$= \hat{M}_{i+1}/(\hat{M}_i - m_i + R_i)$

$\hat{B}_i$ = number of animals that enter the population between the ith and $(i+1)$th samples and which survive until the $(i+1)$th sampling occasion.

$= \hat{N}_{i+1} - \Phi_i(\hat{N}_i - n_i + R_i)$

Note that one cannot calculate $\hat{M}$ for the last sample, $\hat{N}$ for the first or last, Φ and $\hat{B}$ for the last two. $\hat{M}_1$ is bound to be zero.

Calculations are eased if laid out systematically:

$\hat{M}_2 = 10 + (143 + 1)14/(80 + 1) = 34.89$
$\hat{M}_3 = 37 + (164 + 1)57/(70 + 1) = 169.46$
$\hat{M}_4 = 56 + (202 + 1)71/(71 + 1) = 256.18$
etc.

$\hat{N}_2 = 34.89(146 + 1)/(10 + 1) = 466.12 = 466$
$\hat{N}_3 = 169.46(169 + 1)/(37 + 1) = 758.11 = 758$
$\hat{N}_4 = 256.18(209 + 1)/(56 + 1) = 943.82 = 944$
etc.

$\Phi_1 = 34.89/(0 - 0 + 54)$ $\qquad = 0.646$
$\Phi_2 = 169.46/(34.88 - 10 + 143)$ $= 1.009$
$\Phi_3 = 256.18/(169.46 - 37 + 164) = 0.864$
etc.

$$\hat{B}_2 = 758.11 - 1.009(466.26 - 146 + 143) = 290.68$$
$$\hat{B}_3 = 943.82 - 0.864(758.11 - 169 + 164) = 293.13$$
etc.

Confidence limits for $\hat{N}_i$

Methods usually presented for calculating confidence limits of Jolly–Seber estimates are inadequate for the commonly encountered sample sizes. The following method, due to Manly, provides better limits.

Calculate a transformation of each $\hat{N}_i$ and the standard error of the transformation:

$$T_i = \log_e \hat{N}_i + 0.5\log_e[0.5 - 3n_i/8\hat{N}_i]$$

$$s_{T_i} = \sqrt{\left(\frac{\hat{N}_i - m_i + R_i + 1}{\hat{M}_i + 1}\right)\left(\frac{1}{r_i + 1} - \frac{1}{R_i + 1}\right) + \frac{1}{m_i + 1} - \frac{1}{n_i + 1}}$$

For example, the transformation for $\hat{N}_2$ is

$$T_2 = \log_e 466.26 + 0.5\log_e[0.5 - 3(146)/8(466.26)] = 5.6643$$

$$s_{T_2} = \sqrt{\left(\frac{34.89 - 10 + 143 + 1}{34.89 + 1}\right)\left(\frac{1}{80 + 1} - \frac{1}{143 + 1}\right) + \frac{1}{10 + 1} - \frac{1}{146 + 1}} = 0.3309$$

Calculate confidence limits for T_i, and their exponents:

$$T_{iL} = T_i - 1.65s_{T_i}; \qquad W_{iL} = e^{T_{iL}}$$
$$T_{iU} = T_i + 2.45s_{T_i}; \qquad W_{iU} = e^{T_{iU}}$$

Continuing the same example:

$$T_{2L} = 5.664 - 1.65(0.331) = 5.118; \qquad W_{2L} = e^{5.118} = 166.98$$
$$T_{2U} = 5.664 + 2.45(0.331) = 6.475; \qquad W_{2U} = e^{6.475} = 648.69$$

95% confidence limits are given by

$$(4W_{iL} + n_i)^2/16W_{iL} \text{ and } (4W_{iU} + n_i)^2/16W_{iU}$$

For the example of $\hat{N}_2$ the limits are

$$[4(166.98) + 146]^2/16(166.98) = 248$$

and

$$[4(648.69) + 146]^2/16(648.69) = 724$$

Goodness-of-fit test

The test is applied to each sample in turn, except the first and the last. All animals caught in the sample are categorised as follows:

f_1 = first captured before this sample, subsequently recaptured
f_2 = first captured before this sample, not subsequently recaptured
f_3 = first captured in this sample, subsequently recaptured
f_4 = first captured in this sample, not subsequently recaptured

Calculate

$$a_1 = f_1 + f_2, \quad a_2 = f_3 + f_4, \quad a_3 = f_1 + f_3, \quad a_4 = f_2 + f_4$$
$$n = f_1 + f_2 + f_3 + f_4$$
$$g_1 = \sum f \log_e f, \quad g_2 = \sum a \log_e a$$
$$G = 2(g_1 - g_2 + n \log_e n)$$

Meadow Voles *Microtus pennsylvanicus* trapped over 6 occasions in Maryland, USA, by J. D. Nichols (Pollock *et al.* 1990) provide an example for this calculation.

For the second sample the figures were

$$f_1 = 47, \quad f_2 = 31, \quad f_3 = 29, \quad f_4 = 14$$
$$a_1 = 78, \quad a_2 = 43, \quad a_3 = 76, \quad a_4 = 45$$
$$n = 121$$
$$g_1 = 47 \log_e 47 + 31 \log_e 31 + 29 \log_e 29 + 14 \log_e 14 = 422.01$$
$$g_2 = 78 \log_e 78 + 43 \log_e 43 + 76 \log_e 76 + 45 \log_e 45 = 1001.99$$
$$n \log_e n = 121 \log_e 121 = 580.29$$
$$G = 2(422.01 - 1001.99 + 580.29) = 0.62$$

This G should be compared with χ^2 with 1 degree of freedom.

The G values for samples 2, 3, 4 and 5 of Nichols's study of voles, with their associated probabilities, are

sample (i)	G	P
2	0.62	0.43
3	7.69	0.005
4	14.55	0.0001
5	1.83	0.17

Thus samples 3 and 4 give evidence that the assumptions are violated.

The sum of G values provides a test of the overall goodness-of-fit, with a number of degrees of freedom equal to the number of samples providing individual G values. In this example the sum is 24.69, with 4 degrees of freedom, which is highly significant ($P < 0.0001$).

Samples for which any of the expected frequencies of the four groups of animals are less than 2 should be left out of this test.

Developments of the method

The basic Jolly–Seber method has given rise to a number of modifications. One can assume, for example, that there are no gains during the period of study, that the survival rate is constant, or that the capture rate is constant. All these produce more precise estimates (at the risk of violating the new assumptions). One can also allow for temporary trap responses or carry out integrated analyses of data for different age classes.

A further important development is applicable to animals that can be studied for an intensive period every year for several years, the study period being one in which gains and losses are negligible (though there are gains and losses between the annual study periods). It essentially combines a series of annual closed population models with a between-year open model to allow better estimates of population size and turnover rates than would be obtained from a simple Jolly–Seber model.

What area does the study cover?

Some populations live in circumscribed habitat islands that are small enough to be surveyed in their entirety by a single mark–recapture study. More commonly, one is faced with setting a grid of traps in a small area that is representative of a much larger area occupied by the species of interest. It is a basic assumption of the mark–recapture idea that the animals are sufficiently mobile for the marked ones to mix thoroughly with the others; if this is true, then it will also be true that animals from the surrounding area will sometimes enter the study area and may be caught. Thus the population being studied includes animals from an ill-defined strip around the grid of traps, as well as those living on the grid itself.

Various methods have been used to estimate the width of this strip. Most are *ad hoc* and difficult to justify, except insofar as they are likely to give an answer that is closer to reality than if one simply assumes that there is no problem of marginal animals. The best method, in contrast, is based on a simple geometrical model; it requires that the grid of traps is square, with the same number of traps in both directions and all traps the same distance apart (though it could be adapted to rectangular, hexagonal, or other shapes). It assumes that the density and mobility of the animals is constant not just over the whole grid but also in the immediately surrounding area. Since the capture probability of animals that are resident in the study area is likely to be greater than that of animals that originate from the surrounding area, the Burnham & Overton method is likely to be the best method to use in this situation.

Simple geometry dictates the relationship between the number of animals estimated to be present in the study area and the surrounding strip (N), the population density (D) and the size of the grid (plus the surrounding strip). If the grid measures $L \times L$ and the width of the strip is W (Figure 2.2) then

$$N = D(L^2 + 4LW + \pi W^2)$$

N is estimated in the mark–recapture study and L is fixed by the investigator; D and W are unknown. Hence, if two grids of different sizes are used, D and W can be estimated. In practice, one uses the entire data set to provide one estimate of N and the data derived from a

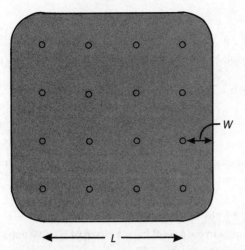

Figure 2.2 A square trapping grid, showing the whole area from which animals caught on the grid may originate, including both the grid itself (which has sides of length L) and a surrounding strip (of width W).

central subset of the grid to provide the other (Box 2.8). The precise size of the central grid chosen is not important; one containing about a quarter of the total number of traps on the grid seems satisfactory. (It is statistically more efficient to use a whole series of nested grids, working down from the entire grid, but this involves a complex analysis that is beyond the scope of this book. The simple method of Box 2.8 will suit most purposes.)

Box 2.8. **Estimating true density from a mark–recapture study based on a square grid**

Basics

$\hat{N}_2$ = estimated number of animals based on the whole grid
$\hat{N}_1$ = estimate based on a central subset of the grid
L_2 = length of the side of the whole grid
L_1 = length of the side of the central subset
$\hat{D}$ = estimate of true density of animals in the area
$\hat{W}$ = estimated width of boundary zones outside the grids, from which animals trapped on the grid may also come

Calculations

$A = \pi(\hat{N}_2 - \hat{N}_1)$
$B = 4(L_1\hat{N}_2 - L_2\hat{N}_1)$

$$C = L_1^2 \hat{N}_2 - L_2^2 \hat{N}_1$$
$$\hat{W} = (-B \pm \sqrt{B^2 - 4AC})/2A$$

(of the two theoretical values of $\hat{W}$, only one will be positive; the negative one is ignored)

$$\hat{D} = \hat{N}_2 / (L_2^2 + 4L_2 W + \pi W^2)$$

Example

Feral House Mice *Mus musculus* trapped on a 10×10 grid on a saltmarsh in California, USA; the traps were at 3-m intervals (Otis *et al.* 1978).

The Burnham & Overton method was used to obtain estimates:

$\hat{N}_2 = 194 =$ number of animals in the area of the main grid of 10×10 traps (an area with sides measuring 9×3 m $= 27$ m $= L_2$)

$\hat{N}_1 = 47 =$ number of animals in the area of the central 4×4 traps (an area with sides measuring 3×3 m $= 9$ m $= L_1$)

Calculation

$$A = \pi(194-47) = 461.81$$
$$B = 4(9 \times 194 - 27 \times 47) = 1908$$
$$C = (9^2 \times 194 - 27^2 \times 47) = -18549$$
$$\sqrt{B^2 - 4AC} = \sqrt{1908^2 - 4(461.81)(-18549)} = 6156.7$$
$$\hat{W} = (-1908 \pm 6156.7)/2(461.81) = 4.60 \text{ m}$$

Thus the animals caught on the grid were coming from an area extending 4.60 m beyond the outermost traps all round.

From the basic expression, density may be estimated as

$$\hat{D} = 194/(27^2 + 4 \times 27 \times 4.60 + 4.60^2 \pi) = 0.150 \text{ animals/m}^2$$

Reference

Otis *et al.* (1978).

Other methods based on trapping and marking

Partial trapping-out: the removal method

If one traps a closed population over a period and removes the trapped animals, then the size of the catches will decline as there are fewer and fewer animals left to be caught. If the numbers caught in each trapping session are plotted on a graph, relative to the numbers

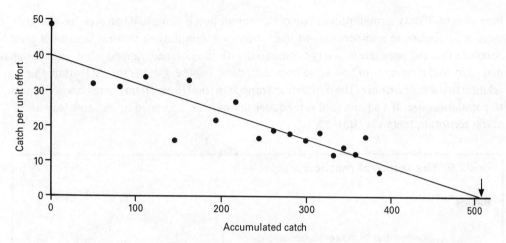

Figure 2.3 The removal method applied to Black Rat *Rattus rattus* catches in 70 houses in Freetown, Sierra Leone, over 18 days. (Data from Leslie & Davis 1939.) The arrow indicates the estimated population size.

already caught in previous trapping sessions, the decline is clear (Figure 2.3). By fitting a line to the data and extrapolating it until it cuts the horizontal axis, one obtains an estimate of the total population (arrow in Figure 2.3): at this point, one's catches would yield no more animals, however long they continued.

Figure 2.3 refers to a study in which the same number of traps was used every day. The trapping effort can be varied from day to day: one plots catch per unit effort on the vertical axis. It is important that trapping effort is measured correctly. For example, doubling the number of traps may not double true trapping effort if neighbouring traps are now so close together that they interfere with each other; or if there are few traps relative to the number of animals and if each trap can catch just one animal at once, then the number caught may be limited by the number of traps.

A major problem with this method is that immigrants may replace the living space vacated by the trapped animals. It is therefore better to mark and release trapped animals rather than remove them, treating the marked animals as 'removed' in the calculations. Any recruitment to the population will result in its initial size being overestimated; losses from the population tend to have the converse effect on the estimate, but not enough to balance out, even when the number of animals lost through death and emigration balances recruitment through birth and immigration. The population estimate will also be incorrect if animals differ in trappability or if trappability increases or decreases with time (e.g. through traps becoming less efficient the more they are used). Violations of the assumptions will lead to the points on the graph forming a curve rather than a straight line. A formal goodness-of-fit test should be applied routinely to test departures from the straight line.

Box 2.9 shows how a precise estimate of population size may be calculated, a goodness-of-fit test applied, and confidence limits of the estimate obtained. As can be seen from the example, the estimates are rather imprecise even when a substantial proportion of the population has been caught. If only a small proportion of the population is caught, it can even happen that there is no decline in numbers caught over successive sampling occasions. One must then conclude that the population is large compared with the number trapped – but one cannot make any useful estimate of how large from such data. Table 2.4 shows the PRP attainable in relation to the proportion of the population trapped in total (over all trapping occasions) and to population size. If a square grid is used, density may be estimated in the same way as for mark recapture methods (Box 2.8).

Box 2.9. **The removal method**

Basics

> S = number of trapping occasions
> y_i = catch per unit effort on the ith trapping occasion
> x_i = total number caught before the ith trapping occasion

Calculate the linear least-squares regression of y on x in the usual way, to obtain

> $\bar{y}, \bar{x}$ = means of y and x
> s_y^2, s_x^2 = variances of y and x
> s_{yx}^2 = covariance of y and x
> b = slope of the regression of y on x

Example

From the data used in Figure 2.3:

> $S = 18$ days, $\bar{y} = 21.8889$, $\bar{x} = 230.5556$
> $s_y^2 = 108.928$, $s_x^2 = 13\,777.882$, $s_{xy}^2 = -1085.817$
> $b = -0.078\,809$

Estimate of total population size

Catchability, $K = -b = 0.078\,809$

Estimated population size, $\hat{N} = \bar{x} + \bar{y}/K = 230.555 + 21.8889/0.078\,809 = 508.302 = 508$

Goodness-of-fit test

Calculate the expected number caught on the ith occasion from

> $E_i = Kf_i(\hat{N} - x_i)$

where f_i = trapping effort on ith occasion (taken as 1.0 when trapping effort is constant). In the study shown in Figure 2.3, 49 animals were caught on day 1 and 32 on day 2 – so that numbers caught before days 1, 2, and 3 were 0, 49 and 81 respectively. Hence:

$$E_1 = 0.078\,809(508.302 - 0) \quad = 40.06$$
$$E_2 = 0.078\,809(508.302 - 49) = 36.20$$
$$E_3 = 0.078\,809(508.302 - 81) = 33.68$$

and so on.

Use these values for the standard goodness-of-fit chi-square test:

$$G = 2\sum O_i \log_e(O_i/E_i)$$

where O_i = actual numbers caught on the ith occasion ($= y_i$, when effort is constant).
For the current data

$$G = 2[49\log_e(49/40.06) + 32\log_e(32/36.20) + 31\log_e(31/33.68)\ldots] = 16.9$$

This G value is to be compared with χ^2 for $S-2$ degrees of freedom (16 in this example).

On this basis, the value of 16.9 is not at all significant. Hence the method can be taken as reliable for these data.

Confidence limits of the estimate

Look up

t = Student's t for $S-2$ degrees of freedom and the $2\frac{1}{2}$% significance level
= 2.467 in this example

Calculate

V = variance about the regression line
$= [s_y^2 - (s_{xy}^2)^2/s_x^2](S-1)/(S-2)$
$= [108.928 - (-1085.817)^2/13\,777.882]17/16 = 24.8160$

Further calculate

$A = K^2 - Vt^2/(S-1)s_x^2$
$ = 0.078\,809^2 - 24.8160(2.467)^2/17(13\,777.882)$
$ = 0.005\,566\,04$
$B = 2\bar{y}K$
$ = 2(21.8889)(0.078\,809) = 3.450\,08$
$C = \bar{y}^2 - Vt^2/n$
$ = (21.8889)^2 - 24.8160(2.467)^2/18 = 470.733$
$W_1,\ W_2 = \{B \pm \sqrt{[B^2 - 4AC]}\}/2A$

$$= \{3.45008 \pm \sqrt{[3.450\,08^2 - 4(0.005\,566\,04)(470.733)]}\}/2(0.005\,566\,04)$$
$$= 203 \text{ and } 417$$

95% confidence limits are

$$\bar{x} + W_1 = 434 \quad \text{and} \quad \bar{x} + W_2 = 648$$

Reference

Seber (1982).

Simultaneous marking and recapture: the method of Wileyto et al.

Suppose that one uses traps of two different sorts in a closed population. One is a normal trap, which catches animals permanently; the other is identical, except that animals that enter it are automatically marked and are able to escape. Such automatic marking is typically carried out by having the inside of the trap coated with a dye that colours the animals that enter. The traps from which the animals cannot escape catch both marked animals (which have been through the other traps) and unmarked animals (which have not). The relative numbers of the two provide an estimate of total population size.

The method assumes equal numbers of the two sorts of traps and that they are equally efficient at catching animals (but that the animals then escape from the marker traps). It works best if the traps are operated in pairs, with one trap of each type in each part of the study area. Differences in trappability between animals do not bias this method badly – but serious bias arises if marked animals are less (or more) likely to enter the permanent traps than are unmarked animals. Although the method applies to closed populations, turnover of the population (through births, deaths, immigration and emigration) has no serious effect so long as the number of animals entering and leaving the population is less than 10% of the numbers being trapped. It is not possible to check the assumptions from the data themselves.

Box 2.10 shows how to calculate the estimate of population size, with approximate confidence limits. Note that the calculations break down if $R \geq U$. In this case $R + U$ provides a reasonable estimate of the population size: it is obvious that this is a minimum estimate; what is less obvious, but true, is that R is only likely to equal U when a very high proportion of the population has been marked. The precision to be expected for this method can be calculated from the expression given for confidence limits in Box 2.10. If one empties the traps periodically, then one may make population estimates as the study progresses and, from their confidence limits, assess for how much longer the work needs to continue.

Box 2.10. **The method of Wileyto *et al*.**

Basics

> U = number of unmarked animals caught in the permanent traps
> R = number of marked animals caught in the permanent traps

Example

Indian Mealmoths *Plodia interpunctella* in a warehouse in the United States (Wileyto *et al.* 1994)

> $U = 74, \quad R = 7$

Estimate of total population size

>   Population size, $\hat{N} = (U + R)^2 / 2(R + 1)$
> $= 81^2 / 2(8) = 410$

Confidence limits of the estimate

Approximate 95% limits are

> $\hat{N} \pm \sqrt{2\hat{N}[U(U+R) + R(U-R)]/R}$
> $= 410 \pm \sqrt{2(410)[74(81) + 7(67)]/7} = 410 \pm 329$
> $= 81$ and 739

Reference

Wileyto *et al.* (1994).

Because animals caught in different traps are not distinguished, the methods of Box 2.7 cannot be used to obtain true density estimates in situations in which animals move into the study area from outside.

Continuous captures and recaptures: the Craig–du Feu method

The previous method involved simultaneous marking and capture, with the relative numbers of marked and unmarked animals being assessed at the end point of the study. Studies in which animals are caught one at a time and quickly released (having been marked if not already carrying a tag) are quite different, but the relative numbers of captures of unmarked and of already marked animals once again provide an estimate of total population size.

Table 2.4. *Percentage of the population that needs to be trapped to achieve a given percentage relative precision (PRP: Box 2.1) in the removal method, in relation to population size (N). (Derived from Zippin 1956)*

Percentages are rounded to the nearest 5.

	Percentage relative precision			
N	60%	40%	20%	10%
200	55	60	75	90
300	50	60	75	85
500	45	55	70	80
1 000	40	45	60	75
10 000	20	25	35	50
100 000	10	15	20	30

Again, the population must be closed – and the estimate cannot be corrected to allow for immigrants.

In practice, one may make estimates of the population size at intervals (or even continuously) during the course of the study. At first, successive estimates fluctuate wildly but they then settle down to more stable values (Figure 2.4) – unless the population is not closed, in which case they tend to drift upwards (or downwards, if animals lost from the population are not replaced by new recruits).

If the animals are individually marked, it is possible to count the number of times that each has been caught and thus use the goodness-of-fit to the zero-truncated Poisson distribution (Box 2.4) as a test for the assumption of equal trappability. (The accuracy of the estimate is probably not much reduced by modest differences in trappability.)

Box 2.11 shows how to estimate population size at any stage during the study, with approximate confidence limits. The expression given for the latter can be used to estimate the precision attainable using this method. Note that it performs better than partial trapping-out, in terms of the precision achieved for a given number of captures. Because estimates and confidence limits may be calculated as the study proceeds, one can (in principle) carry on until a desired level of precision is achieved.

Trapping webs

A trapping web is a set of concentric circles of traps; each circle has the same number of traps in it (usually arranged for convenience so that the traps in the different circles lie in lines, like the spokes on a wheel – Figure 2.5). Ideally, if the innermost circle has a radius R, succeeding circles have radii of 3R, 5R etc. – i.e. the circles lie a constant distance apart and the inner one has a radius equal to half that distance.

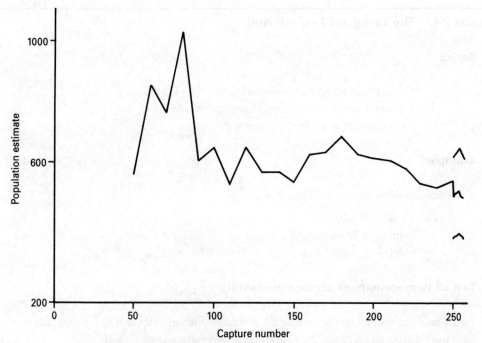

Figure 2.4 The Craig–du Feu method applied to Malachite Sunbirds *Nectarinia famosa* trapped over three days at an isolated food source in South Africa. The line shows successive population estimates in relation to the total number of captures made up to that point, with confidence limits applied to the later estimates. (Data supplied by M. W. Fraser and L. G. Underhill, see Underhill & Fraser 1989.)

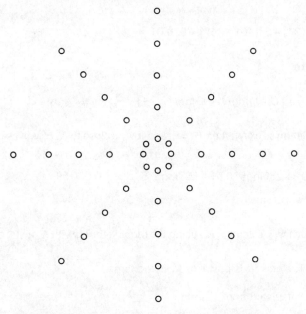

Figure 2.5 Representative layout of trapping web.

Box 2.11. **The Craig–du Feu method**

Basics

> C = total numbers of captures (including all recaptures)
> U = total number of captures of unmarked animals
> f_k = number of animals caught exactly k times

Example

Malachite Sunbirds shown in Figure 2.4.

> $C = 257$, $U = 203$
>
> | number of times caught, | k | 1 | 2 | 3 | 4 |
> |---|---|---|---|---|---|
> | number of birds caught, | f_k | 159 | 34 | 10 | 0 |

Test of homogeneity of capture probability

Before analysis, the test shown in Box 2.4 should be applied. If there is significant departure from the Poisson distribution, the Craig–du Feu estimate will be biased.

For the Sunbird data, the test gives $G = 1.78$ with 2 degrees of freedom, which is not significant.

Preliminary (approximate) estimate of population size

$$\hat{N}_0 = C^2 / [\sum(k^2 f_k) - C]$$
$$= 257^2 / [(1^2(159) + 2^2(34) + 3^2(10) - 257] = 516.01$$

Refinement of estimate

Calculate $H_0 = \log(\hat{N}_0 - U) + (C-1)\log\hat{N}_0 - C\log(\hat{N}_0 - 1)$

If $\hat{N}_0$ equals the true population size (N), then $H_0 = 0$; and if $\hat{N}_0$ is close to N, H_0 will be close to zero. For the Sunbird data:

$$H_0 = \log 313.01 + 256\log 516.01 - 257\log 515.01 = -0.000\,59$$

Since H_0 is negative, $\hat{N}_0$ is too small.

Now guess at another value ($\hat{N}_1$); calculate H_1, using the formula for H_0 with $\hat{N}_1$ instead of $\hat{N}_0$.

If $\hat{N}_1 = 530$, $H_1 = +0.0011$; since H_1 is positive, $\hat{N}_1$ is too large.

The next estimate, $\hat{N}_2$, is calculated from

$$\hat{N}_2 = \hat{N}_1 - (\hat{N}_1 - \hat{N}_0) H_1 / (H_1 - H_0)$$
$$= 530 - (530 - 516.01)0.0011/(0.0011 + 0.000\,59)$$
$$= 520.89$$

$H_2 = +0.000\,12$, calculated as above; since it is positive, $\hat{N}_2$ is also too large.

$\hat{N}_3$ is calculated in the same way as $\hat{N}_2$ (substituting $\hat{N}_1$ and $\hat{N}_0$ by $\hat{N}_2$ and $\hat{N}_1$ respectively)

$$\hat{N}_3 = 520.78$$

$H_3 = -0.000\,0015$; since H_3 is negative $\hat{N}_3$ is too small.

Thus N lies between 520.78 and 520.89. The best estimate of N is thus 521 (to the nearest whole number).

Confidence limits of the estimate

If $\hat{N}$ is our best estimate of N, its 95% confidence limits are

$$\hat{N} \pm 2\sqrt{\hat{N} \Big/ \left(e^{C/\hat{N}} - \frac{C}{\hat{N}} - 1\right)} = 521 \pm 2\sqrt{521/(e^{0.4933} - 0.4933 - 1)} = 521 \pm 120 = 401 \text{ and } 64$$

References

Craig (1953).
du Feu *et al.* (1983).

Animals are caught over a series of trapping sessions, until no more new animals are being caught in the central traps (Table 2.5). It is likely that new animals will still be getting caught in the outer traps (Table 2.5) since, although there is the same number of traps in the outer circles, the area of ground from which they draw animals is larger. The total numbers of animals caught in the few circles that are furthest out may be clearly greater than the numbers caught in slightly less marginal circles (Table 2.4, Figure 2.6), as a result of catching animals originating from beyond the outermost circle. If such immigration is obvious from the data, the affected circles should be ignored: for the data in Figure 2.6, for example, it would be appropriate to truncate the data beyond the 18th circle. Undetected immigrants will result in the density of animals being overestimated. For this reason, it is good practice to release trapped animals (having marked them, to ensure that they are not scored again if they are retrapped later); removing animals leaves a population vacuum into which immigrants may move.

The estimation of population density from trapping webs rests on the fact that animals that are near the centre of a web are more likely to be caught than those near (though within) its outer margins – simply because the traps are more crowded in the centre. It might appear

Table 2.5. *Number of* Peromyscus *mice caught in each of successive circles of traps in a trapping web of 20 circles (each with 16 traps) in New Mexico, USA, over four nights.* (*Data from Anderson* et al. *1983.*)

Circle number 1 was the central circle, number 20 the outermost. The distances given are those from the centre of the web to the circle of traps. Note that in this case the distance from the centre point to the first circle was equal to the distance between successive circles, not half that distance (as would be ideal).

Circle number	Distance (m)	Number of unmarked animals captured				
		Night 1	Night 2	Night 3	Night 4	Total
1	3	1	0	0	0	1
2	6	1	0	0	0	1
3	9	0	0	0	0	0
4	12	2	3	1	0	6
5	15	2	0	0	0	2
6	18	0	1	1	0	2
7	21	3	0	0	0	3
8	24	1	0	0	1	2
9	27	3	1	0	0	4
10	30	5	2	0	0	7
11	33	1	2	1	0	4
12	36	5	0	0	0	5
13	39	4	1	2	1	8
14	42	4	2	0	0	6
15	45	3	3	1	0	7
16	48	3	2	1	0	6
17	51	6	1	0	0	7
18	54	3	0	1	1	5
19	57	5	1	2	1	9
20	60	8	2	2	1	13

possible to work out how the probability of capture varies with distance from the centre, from the areas of the successive circles. The central traps are so close together, however, that they interfere with each other (to an unknown degree), upsetting such simple calculations. Since the size of the interference effect is unknown, the variation in probability of capture must be modelled empirically. It is best to test the fit of various models using a special computer program, but here we present just one method, for which the calculations are relatively simple (Box 2.12). Even for this method, the calculation of confidence limits is not straightforward and so has not been included in Box 2.12. In practice, the best way to estimate precision is to operate more than one trapping web, randomly placed in the study

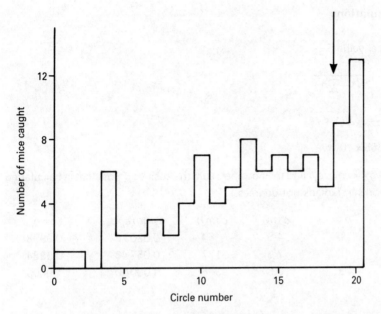

Figure 2.6 Total captures of *Peromyscus* mice in different circles of a trapping web (see Table 2.5). Circle number 1 is the central circle, 20 is the outermost. The arrow marks the point at which the data were truncated for analysis.

area. The mean of the density estimates is then the best estimate of the mean density in the study area; its standard error can be calculated in the usual way.

Box 2.12. Trapping webs

Basics

d_i = distance from the centre of the web to the circle lying exactly halfway between circle i and circle $i+1$

c_i = area enclosed by that circle
    = πd_i^2

c_T = area enclosed by the circle beyond which the data were truncated (see text)

n_i = total number of unmarked animals caught by traps in circle i

Example

The *Peromyscus* data from Table 2.4. The data were truncated beyond the 18th circle of traps, so $c_T = c_{18} = 9677 \, \text{m}^2$.

Preliminary calculations

Calculate the critical value

$$V = \frac{1}{c_T}\sqrt{2/(1+\sum n_i)}$$

$$= \frac{1}{9677}\sqrt{2/(1+76)}$$

$$= 1.665 \times 10^{-5}$$

Calculate values of $g_i(=\pi n_i c_i/c_T)$ and their cosines; sum the cosines. (Note that in calculating the cosines one works in radians not degrees.)

i	n_i	d_i(m)	c_i(m²)	$g_i = \pi n_i c_i/c_T$	$\cos g_i$
1	1	4.5	64	0.020 78	0.9998
2	1	7.5	177	0.057 46	0.9984
3	0	10.5	346	0.000 00	1.0000
.		.	.		
.		.	.		
.			.		
18	5	55.5	9677	15.707 96	−1.0000
	⎯				⎯⎯⎯
	76				4.9426

Calculate

$$a_1 = 2\left(\sum \cos g_i\right)/(c_T \sum n_i) = 2(4.9426)/(9677 \times 76) = 1.344 \times 10^{-5}$$

Is the absolute value (i.e. the value ignoring any minus sign) of a_1 less than the critical value V? In this case it is, so one calculates

$$f = a_1 + 1/c_T$$

$$= 1.344 \times 10^{-5} + 1/9677 = 1.17 \times 10^{-4}$$

If the absolute value of a_1 had been greater than V, one would have calculated values of $\cos 2g$, summed them, and calculated

$$a_2 = 2\left(\sum \cos 2g\right)/c_T \sum n_i$$

If the absolute value of a_2 was less than V, one would have proceeded exactly as above. If it was greater, one would calculate values of $\cos 3g$, to obtain a_3 in the same way as a_1 and a_2 had been obtained. If its absolute value was less than V, one would calculate

$$f = a_1 + a_2 + 1/c_T$$

If a_3 was larger than V, one would calculate a_4, a_5, a_6 in turn, stopping when one reached a value smaller than V. The sum of all the a values except that last one, plus $1/c_T$, is taken as f.

If even a_6 is not less than V, it is usually safest to stop at that point and use

$$f = a_1 + a_2 + a_3 + a_4 + a_5 + a_6 - 1/c_T$$

Estimate of population density

$$\hat{D} = f \sum n_i$$
$$= 1.17 \times 10^{-4} \times 76 = 89 \times 10^{-4} \text{ mice/m}^2$$
$$= 89 \text{ mice/ha.}$$

Test of goodness-of-fit of the model

The model presented here is a Fourier-series model. From this, the expected number of animals caught in each ring of the traps can be calculated and a standard goodness-of-fit test applied.

Begin by calculating $\sum a_k/k$ for all values of k from 1 to m (m being the number of values used in the calculation of f, above). Then extend the table of calculations used to find $\sum \cos g_i$ by tabulating the following:

$$x_i = \pi c_i/c_T$$
$$\sin x_i$$
$$\sin x_i - \sin x_{i-1} \quad \text{(when } i=1 \text{ this is simply } \sin x_i\text{)}$$
$$y_i = \left(\sum \frac{a_k}{k} \right)(\sin x_i - \sin x_{i-1})/\pi$$
$$z_i = (c_i - c_{i-1})/c_T \quad \text{(when } i=1 \text{ this is simply } c_i/c_T\text{)}$$
$$E_i = (y_i + z_i) \sum n_i$$

The values of E_i are the expected numbers of animals caught in each zone.

The calculations are carried out for all values of i except the largest: this value of E_i must be estimated from the difference between the total number of animals caught ($\sum n_i$) and the sum of all the other E_i values.

In the *Peromyscus* example, the first few values are

i	x_i	$\sin x_i$	$\sin x_i - \sin x_{i-1}$	y_i	z_i	E_i
1	0.0207	0.0207	0.0207	0.0009	0.0066	0.56
2	0.0574	0.0573	0.0367	0.0015	0.0117	1.00
3	0.1124	0.1122	0.0549	0.0023	0.0175	1.51

To carry out the goodness-of-fit test, observed and expected values for adjacent zones are combined as much as necessary to avoid expected values less than 5. For the *Peromyscus* data

Zones	O_j	E_j
1–4	8	5.08
5–6	4	5.49
7–8	5	7.41
9–10	11	9.17
11	4	5.17
12	5	5.50
13	8	5.79
14	6	6.03
15	7	6.25
16	6	6.46
17	7	6.69
18	5	6.96

As usual

$$G = 2\sum O_j \log_e(O_j/E_j)$$
$$= 4.93 \quad \text{with 17 degrees of freedom in this case.}$$

This value is not significant when compared with the tabulated values of χ^2, so we can conclude that the Fourier model is an adequate fit to the data.

Reference

Anderson *et al.* (1983).

Whatever statistical model is used, three basic assumptions underlie the analysis: that all animals in the central area are caught; that population density is constant across the web; and that the trappability of the animals is the same in all of the trapping sessions.

Point and line transects

General principles

Suppose that one is observing big game from a low-flying aircraft, whales from a survey ship, or birds from a position in the middle of a forest. Even an experienced observer will not detect animals that are far away from the survey line. Point and line transect methodology may be applied to such situations. It requires that the distances from the observation point or line to each animal are recorded, so it is often known as 'distance sampling'; one then uses a model of the way in which detectability drops off with increasing distance to estimate the number of animals that one has missed. Statistically, point and line transects are analogous to trapping webs (for which trappability declines with distance from the centre point). They are, however, particularly suitable for animals that are difficult to catch, especially those, such as birds, that can be observed directly in the field but only detected with any certainty if they are close. Like trapping webs, transect methods provide direct estimates of density.

The theory of transect methods assumes that all animals at the observation point or on the line are detected. (If they are not, such as in aerial surveys when animals immediately beneath the aircraft cannot be seen, then special models may be used but the estimates these provide are very sensitive to the exact statistical model used for the analysis.) If animals flee away from (or are attracted to) the observer, they must be detected before they are disturbed. (Dealing with the problem of animals fleeing from the observer's path before they are detected presents similar problems to failing to detect the animals on the path.) Population density is assumed to be constant around the observation point or line. If detections of different animals are not independent, as for example if neighbouring birds tend to sing at the same time, the variability of the results is increased. Another example of non-independence would be when animals occur in flocks. In this case, one can treat the flock, rather than the individual, as the sample unit; having estimated the density of flocks, one multiplies by average flock size to estimate the density of the population; unfortunately this simple approach will be upset if detectability is greater for larger flocks, as it usually will be.

Distance sampling methods also depend for their success on the accuracy with which the distance to each animal is measured. In reality, the distances usually have to be estimated. For this reason, it may be better simply to record whether the animal is within or beyond some fixed distance, for it is easier to train observers to estimate one fixed distance than to estimate all distances accurately. This may more than compensate for the fact that methods based on only two recording zones (within and beyond the fixed distance) are inherently less precise than those based on full distance measurements, because they use much less information. As with trapping webs, the best way to estimate overall precision is to undertake a number of transects within the study area, estimating their mean and standard error in the usual way.

Buckland *et al.* (1993) discuss distance sampling methods in detail.

Point transects

A point transect entails the observer remaining at one point for a fixed time and recording the animals he or she sees (or otherwise detects). Distances may be recorded in terms of concentric zones around the point (e.g. up to 50 m, 50–100 m, 100–200 m, etc.) up to some limit beyond which the animals are not identifiable. (An alternative limit is one that excludes about 5% of the animals seen.) One then simply applies the methods of Box 2.12, taking the areas (c_i) as the areas enclosed by the outer limits of each zone.

Alternatively, the exact distance to each animal (d_i) may be measured or estimated. One then calculates the area enclosed by a circle of radius d_i and uses this as the relevant c_i value in Box 2.12. Thus each observation has its own c_i value. Corresponding g values are calculated, based on sample sizes (n_i) of 1. The value of c_T used in this method is the area enclosed by the outer limit that one has imposed on one's observations. Because point transects involve calculating areas from the distances measured, they are especially sensitive to errors in the distances, since the error is squared in calculating the area.

Point transect methodology assumes that there is no immigration into the area during the observation period; otherwise density will be overestimated. The recording period must, however, be long enough for all animals close to the observer to be detected. Indeed, because birds, for example, may fall silent as an observer approaches, it is common to allow a brief 'settling down' period after arriving at the observation point before beginning the observations.

The number of animals recorded at a single point is often small. This problem may be overcome by combining the data from a number of replicate counts made within an area in order to estimate mean density in the area – though one must remember to divide by the number of replicates.

Calculations based on the simple method of only recording how many animals are detected within a fixed distance and how many beyond that distance are presented in Box 2.13. They are based not on the Fourier series (used in Box 2.12) but on the simple model that detectability declines exponentially with distance (i.e. as e^{-kd^2}, were d is distance and k is a constant). Goodness-of-fit cannot be tested for data based on only two recording zones.

Box 2.13. **Point transects with only two recording zones**

Basics

r = radius of first zone (the second extends from r to infinity)
n_1 = number of animals counted within r
n_2 = number of animals counted beyond r
m = number of replicate points in the set

Example

326 replicate counts of 5 min each in conifer plantations in Wales yielded 421 Willow Warblers *Phylloscopus trochilus* within 30 m and 504 beyond that distance (Bibby *et al.* 1985).

Calculation

$$\text{Density} = \frac{n_1 + n_2}{\pi r^2 m} \log_e \left(\frac{n_1 + n_2}{n_2} \right)$$

$$= \frac{925}{\pi (30^2) 326} \log_e \left(\frac{925}{504} \right) = 6.09 \times 10^{-4} \text{ birds/m}^2$$

$$= 6.09 \text{ birds/ha}$$

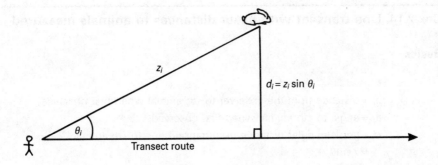

Figure 2.7 Calculating the perpendicular distance from an animal to the transect line.

Line transects

Point transects may waste a lot of time: one may spend time waiting at the point long after all the detectable animals have been observed and one spends time walking from one point to the next. It is often possible to record animals observed while moving. In this case, line transects are to be preferred. They are often used for birds (especially in open habitats, where it is relatively easy to record birds while walking), for aerial surveys of big game, and for ship surveys of whales. The principles and assumptions are the same as those for point transects, though line transects are less sensitive to errors in distance estimates since the calculations depend on the distances themselves rather than their squares. The animals may be recorded in a series of bands on each side of the transect line (corresponding to circles around transect points) or exact distances may be measured. The latter must be the distances perpendicular to the transect line; if animals flee as the observer approaches one should measure (or estimate) both the distance from the observer to the animal when it is first detected and the angle between the animal and the transect, so that the perpendicular distance may be estimated (Figure 2.7). Whether one uses recording bands or exact distances, one should truncate the observations beyond a certain distance; as with points, this should either exclude about 5% of the animals seen or be the limit beyond which animals cannot be identified. Animals are usually recorded on both sides of the line but may be recorded on only one.

Note that line transects are not the same as strip transects, in which one counts all the organisms in a strip across the study area, assuming that every individual in the strip is detected and including none from beyond the edges of the strip. Nor are they the same as line intercepts, in which one counts the number of sessile organisms crossed by a line laid down across the study area. (If one measures the length of line intercepted by the organisms, this provides a good measure of the area covered by that species.)

Estimating densities from line transects is similar to estimating from point transects but uses distances rather than area. Box 2.14 gives an example, based on a very small sample. If one examines the data in this example (especially if one plots a frequency histogram of the

Box 2.14. **Line transect with exact distances to animals measured**

Basics

z_i = distance from the observer to ith animal when first observed
θ_i = angle to ith animal when first observed
d_i = perpendicular distance from transect line to ith animal
    = $z_i\sin\theta_i$
d_T = distance beyond which data were truncated
l = length of transect

Example

10 km transect in the Sudan, to survey White-eared Kob *Kobus kob* (Krebs 1989). Distances beyond 120 m were truncated (see text): i.e. $d_T = 120$.

The formula for g_i values is as in Box 2.11, but the line transect method uses distances, not areas:

$$g_i = \pi n_i d_i / d_T$$

Note that where the data are a list of individual observations (rather than numbers of animals in a series of recording zones) all values of n_i equal 1.

i	z_i (m)	θ_i	d_i (m)	g_i	$\cos g_i$
1	150	38	92	2.409	−0.743
2	200	55	164 x	—	—
3	160	8	22	0.576	0.839
4	200	17	58	1.518	0.052
5	250	39	157 x	—	—
6	130	42	87	2.278	−0.649
7	150	10	26	0.681	0.777
8	130	23	51	1.335	0.233
9	200	55	164 x	—	—
10	100	46	72	1.885	−0.309
11	140	31	72	1.885	−0.309
12	200	25	85	2.225	−0.609

$$ \overline{}$$
$$ -0.718$$

$$a_1 = 2(\textstyle\sum \cos g_i/(d_T \sum n_i) = 2(-0.718)/120 \times 9) = -0.001\,33$$
$$V = \frac{1}{d_T}\sqrt{2/(1+\textstyle\sum n_i)} = \frac{1}{120}\sqrt{2/(1+9)} = 0.003\,73$$

Since the absolute value of a_1 is less than V, one calculates

$$f = a_1 + 1/d_T$$
$$= -0.001\,33 + 1/120 = 0.007\,00$$

Had the absolute value of a_1 been greater than V, one would have proceeded as in Box 2.11, calculating further a values as required in order to obtain f.

Estimate of population density

For two-sided line transects, the density estimate must be adjusted by twice the transect length:

$$\hat{D} = \sum n_i/2l$$
$$= (0.007\,00 \times 9)/(2 \times 10\,000) = 3.2 \times 10^{-6}\ \text{kob/m}^2$$
$$= 3.2\ \text{kob/km}^2$$

distances, d_i), one sees that there is a suspicious clustering of three values (25% of the data!) around 160 m, the next largest value being 92 m. This suggests that some distances in the range 100–150 m have been overestimated and some distances greater than 150 m underestimated. An empirical solution, applied here, is to exclude all values of d_i greater than 120 m (marked x in Box 2.14). Note the value of plotting a frequency histogram of the distances; one expects high frequencies of short distances, then a rapid falling away through the middle distances, with a long tail of a few large distances. As with the trapping web (Figure 2.6), the histogram allows one to truncate the tail of the distribution if it appears to contain too many observations. It also allows one to check that animals close to the transect line are not fleeing before being detected, since this would result in a dip in frequency of the shortest distances.

Box 2.15 presents calculations based on just two recording bands, similar to the corresponding approach for point transects. Again, the goodness-of-fit of this method cannot be tested.

Migration counts

Grey-Whales *Eschrichtius robustus* migrating from Mexican waters to the Arctic pass close inshore off the Californian coast and may be well observed at a number of points. They can be counted and the distances at which they are passing measured. Since this is the equivalent of a one-sided line transect (merely with the observer fixed and the animals moving, rather than the converse!), the density of animals per hour can be estimated. Watches throughout the migration season allow an estimate of the total number of animals. The method can be applied to similar situations elsewhere.

Box 2.15. **Line transects with only two recording bands**

Basics

r = distance from the transect line to the boundary between the two zones (the second extends to infinity)
n_1 = number of animals counted within r
n_2 = number of animals counted beyond r
l = length of transect

Example

Skylarks *Alauda arvensis* counted on a 2-km transect of arable farmland in Norfolk, England, yielded 10 birds within 100 m of the transect line, 14 beyond (own data).

Calculation

$$\text{Density} = \frac{n_1 + n_2}{2rl} \log_e \left(\frac{n_1 + n_2}{n_2} \right)$$

$$= \frac{24}{2 \times 100 \times 2000} \log_e \left(\frac{24}{14} \right) = 3.2 \times 10^{-5} \text{ birds/m}^2$$

$$= 32 \text{ birds/km}^2$$

Note: in this example, birds were recorded on both sides of the line; for one-sided transects, do not divide by 2 in the estimation of density.

Plotless sampling

These methods, discussed in detail by Diggle (1983), are suitable for sessile organisms that are so sparsely distributed that laying down sampling areas and counting all the individuals in each is inordinately time-consuming. Censusing the trees of one of the less common species in a forest would be an example.

The general idea is either to pick *m* individuals at random and to measure the distance from each to its nearest neighbour or to go to *m* random points and measure the distance from each to the nearest individual of the species in question. If the individuals are randomly distibuted, density of the population can be estimated from such measurements. Unfortunately, animals and plants are rarely randomly distributed and these methods are seriously biased if the distribution tends to be either aggregated or more regular than random. The biases are, however, in opposite directions (if aggregated, nearest-neighbour overestimates density,

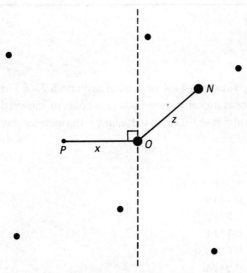

Figure 2.8 The *T*-square method. The black circles represent individuals of the study species. *O* is the nearest individual to a random point *P*; *N* is *O*'s nearest neighbour on the opposite side of the dashed line (which is perpendicular to the line *OP*). For further explanation, see text.

point-to-object underestimates it; if regular, the converse is true). This makes *T*-square sampling, which combines both types of measurement, useful.

Figure 2.8 illustrates the procedure. Choose a random point (*P*); measure the distance (*x*) from the point to the nearest individual of the species in question (*O*); lay down a line at right angles to the line joining *O* and *P* (the dashed line in the figure); measure the distance (*z*) from *O* to the nearest individual (*N*) on the opposite side of the line to *P*. Make a number of such measurements – 10 is the recommended minimum – each based on a new random point.

The estimation of density from these measurements is straightforward (Box 2.16). Although, as explained above, the method is reasonably robust in the face of non-random distributions, Box 2.16 illustrates a test of randomness which it is prudent to apply routinely, so that potential biases are not forgotten. Other plotless sampling methods are considered in Chapter 3.

It is important for plotless sampling that the core area in which one places the *m* random points should be somewhat smaller than the total study area, to allow for the possibility that the nearest individuals to some of the points may lie outside the core area.

Box 2.16. *T*-square sampling

Example

Japanese Black Pine *Pinus thunbergii* in a core 4×4-m quadrat, within a 5.7×5.7-m quadrat (Numata 1961). Ten ($=m$) pairs of measurements were made (x_i = point-to-sapling distances, z_i = nearest-neighbour distances, all in metres). Calculated values in the third column are used to test for randomness of the distribution.

x_i	z_i	$x_i^2/(x_i^2 + z_i^2/2)$
0.1	0.3	0.1818
0.3	0.2	0.8182
0.1	0.4	0.1111
0.8	0.5	0.8366
0.3	0.1	0.9474
0.7	0.7	0.6667
0.6	0.7	0.5950
0.9	0.7	0.7678
0.7	0.3	0.9159
0.5	0.3	0.8475
——	——	——
5.0	4.2	6.6880

Estimation of density

$$\hat{D} = m^2/(2.828 \sum x_i \sum y_i)$$
$$= 10^2/[2.828(5.0)(4.2)] = 1.68 \text{ trees/m}^2$$

(This compares well with 1.625 obtained from a complete census.)

Test for random distribution

Calculate

$$t' = \{\sum[x_i^2/(x_i^2 + z_i^2/2)] - m/2\}\sqrt{(12/m)}$$
$$= \{6.6880 - 10/2\}\sqrt{12/10} = 1.85$$

If t' is greater than $+1.96$, the distribution is significantly more regular than a random distribution; if it is less than -1.96, it is significantly clumped. In either case, $\hat{D}$ may be biased.

Population indices

The idea of an index

The ideal index

An index is a measurement that is related to the actual total number of animals or plants. Ideally, the relationship should be such that the ratio of the index to numbers is constant:

$$I/N = K$$

In this case, even if the index ratio (K) is unknown, one can compare populations in different places or at different times by comparing the indices. In reality, the index ratio tends to vary, as can be appreciated by considering some examples of indices.

Examples of indices

The number of animals seen during a certain time period while standing at a point or the number seen while walking a set distance (that is, point and line counts rather than point and line transects, which require distances to be measured so that actual densities can be estimated) are merely indices; they vary according to the competence of the observer, the behaviour of the animals (they may be less detectable when resting, for example), and all those environmental factors that influence detectability, such as the density of the vegetation, the lie of the ground, and weather conditions. The number of calls heard per minute can be an index of the numbers of frogs in an area but calling rate varies with season, time of day (or night), and weather. The number of animals shot by hunters is commonly used to index the numbers of game animals; it is influenced by the number and enthusiasm of hunters. The number of animals caught in pitfall traps depends not only on the density of animals in the area but on their activity (which varies with season, weather, etc.), and on the surrounding vegetation, on the colour of the trap, on the size and shape of the aperture, on the material, depth and shape of the trap (which influence whether animals can escape), on whether or not there is a preservative fluid in the traps, what that fluid is and how deep it is (which influence the attractability of the traps and the ease of escape from them), and on the spacing and arrangement of the traps. The population of small seeds or micro-organisms in soil may be assessed by germinating or culturing them; the numbers observed will depend crucially on the culture conditions. Tracks in mud provide an index of deer populations, which is influenced both by the animals' activity and by soil conditions; tracks left by small mammals on smoked paper left in tunnels provide more standardised recording conditions but still depend on activity as well as numbers. The number or weight of droppings in an area depends both on the size and metabolic activity of animals as well as their numbers.(This can be turned to advantage by ecologists studying, for example, the availability of caterpillar food to birds, since actively feeding caterpillars are easier for birds to find than inactive ones and large ones provide more food than small ones; thus the weight of caterpillar droppings falling on the forest floor may be a better index of available food supply than is a simple count of caterpillar numbers.)

Some indices refer only to part of the population. Thus a count of breeding territories is an index only of the breeding population. Even for the breeders, it may be an imperfect index, for in some species not all the breeders maintain territories, while in others individual males may have two or more separate territories at the same time. If unoccupied nests can be recognised, the number of breeding nests is the same as the number of breeding females in many species. Indices that are so closely tied to numbers can be used instead of counts of the animals themselves.

Counts and density estimates as indices

The first example of an index listed above was an incomplete direct count. Earlier in this chapter, we have considered various indirect methods of estimating population numbers and density, noting that it is often difficult to avoid bias. Such biased estimates can be taken as indices, recognising that the factors leading to the bias should not vary if the index ratio itself is to be constant, as is ideal. We have also seen that it may not be possible to estimate the true density of a population using many of the techniques based on trapping, because one is not sure from how large an area the trapped animals have been drawn; the estimated size of the trappable population is then an index of population density in the area.

Overcoming variation in the index ratio

The value of standardisation

Much of the variation likely to affect index ratios can be substantially reduced by standardisation – making the observations under similar weather conditions, at the same time of day, and in the same season; using the same type, number and layout of traps (containing the same attractants or foods); having the observations made by the same person. Simultaneous trapping sessions in different places ensures standardisation of factors such as weather, time of day, and season – though subtle microclimatic differences between the sites can cause different levels of activity in their populations. Standardisation is essential if one wishes, as is usual, to compare populations in different places or to study the variation in numbers over time. More generally, it is valuable in reducing the variation in one's measurements.

The value of multiple, randomised observations

Suppose that an observer wished to compare the number of animals in two areas, knowing that the bias in her counts was strongly weather dependent. If she made counts in the two areas on two consecutive days, she would have no way of knowing whether the differences between the counts were the result of differences in the populations or of differences between the weather on the two days. To overcome this problem, she would need to extend the study over several days; on half the days (chosen at random, see p. 75) she would count in one area and on the other days she would count in the second area. The standard techniques of analysis of variance could then be used to compare the mean counts in the two areas, the effects of weather being subsumed in the error variance.

Allowing for variation

Better than simply including the effects of weather (or any other factor that influences the population index) in the error variance, one can take them into account in the analysis. An imaginary example is shown in Box 2.17, in which the variable to be allowed for is the time of day. The method assumes that the effect of time of day is the same in the different study areas; if it is not, it is more difficult to allow for it. In the example in Box 2.17, the 95% confidence limits of the difference between the mean indices in the two areas are +0.6 and +35.2 (in favour of Area B). If time of day had been ignored in this analysis (so that the data for all three times were lumped into a single set for each area), the variation associated with time would have been subsumed in the residual variance, so the mean squares would have been

Place   2412.03 (1 d.f.)
Residual   2245.39 (24 d.f.)

As a result, the confidence limits of the difference between the mean indices of the two areas would have been much wider (−17.6 and +53.4). Taking extraneous factors into account thus increases the precision with which means and differences are estimated.

Had the example in Box 2.17 involved more than two areas, the confidence limits of the difference between any two of them would have been worked out in the same way, using the residual mean square from an overall analysis of all the areas.

Correction factors

The mean indices for the three times of day in Box 2.17 are 71.8 (morning), 139.8 (mid-day), and 169.3 (afternoon). Since these observations have established the differences between these times, one could, in principle, take observations made at only one time of day from other areas and correct them to some standard time of day. This practice is, however, best avoided, as it assumes that the differences between times of day are the same in all the other areas, which may not be true.

Regression techniques

Suppose one was trying to index frog populations by counting the number of calls heard in standard ten-minute periods, at a standard time of night. It is likely that calling rate varies with temperature, so it would be best to make observations only on nights on which the temperature fell within a certain range. Were this to be impossible, one could classify the nights into categories such as 'cool', 'warm', and 'hot' and then apply the analysis of variance method of Box 2.17 to the data. Greater precision would be obtained if one measured the temperature at the time of each observation and used the usual statistical techniques of regression analysis to discover the relationship between call rate and temperature. The extent to which the call rate at an individual location departed from the regression line would then provide an index corrected for the effect of temperature (Figure 2.9).

Box 2.17. **Using analysis of variance to allow for extraneous sources of variation in index studies**

Imaginary example

An index that is greatly affected by time of day. Observations were made to compare populations in two areas; because not all observations could be made at a standard time of day, this factor has to be allowed for in the analysis.

The observations were as follows:

	morning	mid-day	afternoon
Area A	40, 49, 66, 69, 90	100, 107, 130, 140, 174	140, 150, 165, 170, 180
Area B	59, 69, 84, 87, 105	119, 125, 145, 163, 195	150, 170, 170, 195, 203

Analysis of variance

The detailed methods used to calculate the mean squares are described in statistical texts and so are not repeated here.

Source of variation	Mean squares	degrees of freedom
Time	25 000.53	2
Place	2 412.03	1
Interaction	4.22	2
Residual	535.87	24

Are the effects of time the same in the two areas? This hypothesis is tested by the variance ratio for the interaction:

$$F = 4.22/535.87 = 0.008, \text{ with 2 and 24 degrees of freedom}$$

This value is not significant, so we can conclude that the effects of time of day are the same in the two areas.

Differences between the areas

Mean indices, averaged over all times of day, are

$$A: (40 + 49 + \ldots 180)/15 = 118.0$$
$$B: (59 + 69 + \ldots 203)/15 = 135.9$$

The mean difference between the two areas is thus

$$d = 135.9 - 118.0 = 17.9$$

The standard error of this difference is

$$s_d = \sqrt{2s^2/n}$$

where

> s^2 = residual mean square from the analysis of variance
> n = number of observations in each of the samples being compared

In this case,

$$s_d = \sqrt{2 \times 535.9/15} = 8.45$$

95% confidence limits are

> difference $\pm t \times s_d$

where t is Student's t for $n-2$ degrees of freedom and the 5% significance level.

The limits in this case are

> $17.9 \pm 2.05 \times 8.45 = +0.6$ and $+35.2$

Cases in which the sample sizes are not the same in all combinations of area and time are more complex to calculate but employ the same principles.

Greater precision is obtained by making several observations at each location, on different nights, so that separate regression lines can be fitted to each (using the standard techniques of analysis of covariance). The differences in elevation of the lines then provide measures of differences between their populations (Figure 2.10).

When the index ratio varies with population density

The ratio of an index to the number of animals present may vary not only in relation to extrinsic factors but in relation to population density itself. For example, the frequency with which an animal calls may increase if others are calling nearby; if so, a doubling in numbers may lead to the call rate increasing more than twofold (Figure 2.11, dashed line). Conversely, an observer may be overwhelmed by the number of calls at high densities, so that changes in the counts are less than proportional to changes in numbers present (Figure 2.11, continuous line).

Indices may, indeed, have maximum possible values. If an index comprises the number of animals caught in a session and if only one animal can be caught per trap, then the index has a maximum value set by the number of traps. Another example is when the index is based on frequency of occurrence: it is sometimes not practical for observers to count the numbers of animals but easy for them to record whether they have seen any animals at all during a particular observation period in a particular area; the proportion of sampling occasions, or

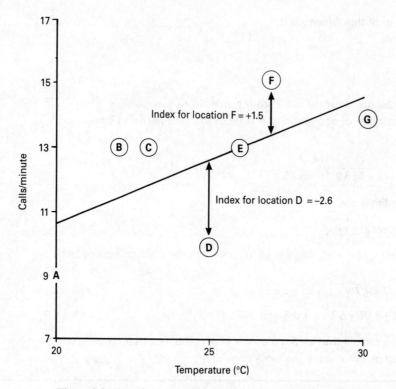

Figure 2.9 Imaginary example showing total call rates of frogs at different locations (A–G) in relation to temperature at the time the observations were made. The deviation of individual points from the regression line provides an index of numbers at that location, corrected for the effect of temperature on call rate.

of sampling areas, in which animals were seen then provides an index of numbers of the species in question; it is not, however, a useful index for abundant and widespread species, which are seen on almost every occasion.

Indices that are not linearly related to numbers are useful provided one bears in mind that a difference of, say, 10 points in the index has different implications if it is the difference between 20 and 30 from the implications it has if it is the difference between 120 and 130. This means, in particular, that average values have to be treated with caution. Box 2.18 shows how samples of observations can have identical mean values of the index but different mean population sizes (compare samples *A* and *B*) and conversely (compare *B* and *C*); indeed, one sample may have a smaller mean population size than another but a larger mean index (compare *C* and *D*). When the relationship between the index and numbers is convex (like the continuous line in Figure 2.11), samples that are more variable (*C* compared with *B*, in Box 2.18, for example) have lower mean indices for a given mean population size. The converse is true for concave relationships (Figure 2.11, dashed line).

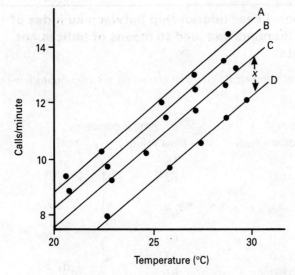

Figure 2.10 Imaginary example showing total call rates of frogs at four different locations (A–D) in relation to temperature at the time at which the observations were made, there being several observations at each location. The differences in elevation (e.g. x) between the parallel regression lines provide indices of differences in numbers, corrected for the effect of temperature on call rate.

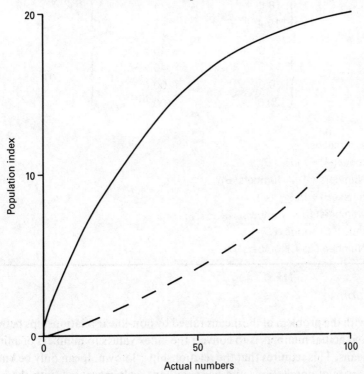

Figure 2.11 Non-linear relationships between index values and number of animals present (imaginary examples). See text for further explanations.

Box 2.18. **Showing how a non-linear relationship between an index of population size and actual numbers can lead to means of indices not reflecting means of numbers**

The relationship between index and numbers used is that shown by the continuous line in Figure 2.11.

		Index values		Actual numbers	
Sample	Mean	Individual values		Individual values	Mean
A	18.0	16.0		50	
		18.0		67	
		18.0		67	71
		20.0		100	
B	18.0	18.0		67	
		18.0		67	
		18.0		67	67
		18.0		67	
C	17.6	15.2		47	
		17.1		60	
		18.6		74	67
		19.4		87	
D	17.0	14.0		40	
		14.0		40	
		20.0		100	70
		20.0		100	

Note that, in terms of means:
1. Although   Index(A) = Index(B),
        Numbers $(A) \neq$ Numbers(B).
2. Although   Index$(B) \neq$ Index (C),
        Numbers(B) = Numbers(C).
3. Although   Index$(C) >$ Index(D),
        Numbers$(C) <$ Numbers(D).

Calibration

One way of dealing with the problem of the means raised by non-linear relationships between population indices and actual numbers is to convert the index values to numbers of animals before calculating means. This requires that the relationship is known. It can only be known if one has studied a series of populations and has simultaneously measured both the index

Box 2.19. **The calibration of an index**

Example

Over a number of different winters and in a number of different areas of northern Wisconsin, USA, the number of trails of White-tailed Deer *Odocoileus virginianus* encountered on strictly standardised transects was counted; the population densities of the deer in these areas were known through more direct methods (McCaffery 1976). Provided that the temperature had dropped to $-2°$ by 20 October, the number of trails was linearly related to the density of deer:

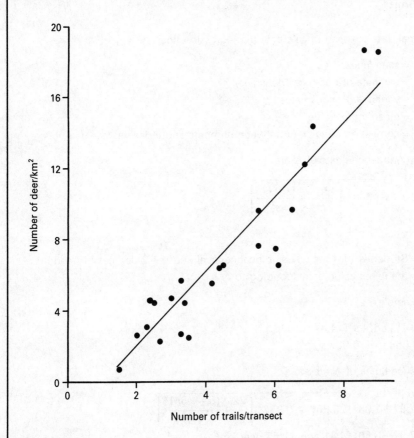

Methods

Calculate the least-squares regression of density (or numbers) on the index in the usual way to obtain the equation

$$N = a + bI$$

For the deer, the relationship between numbers/km^2 (N) and trails/transect (I) (the line plotted on the above graph) was

$$N = 2.16I - 2.83$$

From this equation, one can estimate the density of deer in an area simply from the number of trails/transect. Suppose, for example, that one encountered 5.0 trails per transect; then the estimated number of deer would be

$$\hat{N} = 2.16 \times 5.0 - 2.83 = 7.97 \text{ deer/km}^2$$

Confidence limits

From the original data set used to calculate the regression line:

m = sample size
s_N^2 = variance of numbers (densities)
s_I^2 = variance of index values
$\bar{I}$ = mean index value
r = correlation coefficient between numbers and index values

Then the 95% confidence limits of $\hat{N}$ are

$$\hat{N} \pm t \sqrt{s_N^2 (1 - r^2) \left[\frac{m-1}{m} + \frac{(I' - \bar{I})^2}{s_I^2} \right]}$$

where

t = Student's t for the 5% significance level and $m - 2$ degrees of freedom
I' = the index value corresponding to $\hat{N}$

For the deer example,

$$s_N^2 = 23.86, \quad s_I^2 = 4.47$$
$$\bar{I} = 4.54, \quad r = 0.937$$

Hence, confidence limits of $\hat{N}$ are

$$7.97 \pm 2.08 \sqrt{23.86(1 - 0.937^2) \left[\frac{22}{23} + \frac{(5 - 4.54)^2}{4.47} \right]}$$
$$= 7.97 \pm 3.56 = 4.41 \text{ and } 11.53 \text{ deer/km}^2$$

Warning

Estimates made from calibration equations should be applied only under the same range of conditions and over the same range of densities as the original calibration data.

and the actual numbers. Such calibration then allows numbers to be estimated from studies in which only indices have been obtained.

Box 2.19 presents an example of such a calibration. It is a simple case in which a linear relationship between index and numbers is applicable; non-linear relationships demand more complex calculations.

Two important restrictions apply to calibrations. First, they should not be applied outside the range of population sizes used in the original calculations, as the relationship between index and numbers may not be the same outside that range. Second, they should not be applied to populations living in different habitats, at different times, etc.; again, the relationship may not be constant over all habitats, times, etc. In practice, these restrictions may be eased, if one's knowledge of the species indicates that it is safe to do so; but this should only be done with great caution.

Calibration may also be based on knowledge other than direct comparison of indices and numbers. For example, it is virtually impossible to estimate populations of adult seals; they spend much of their time under water and are difficult to capture. However, for some species, it is possible to establish the number of pups born each year, since they are born on land. The age of seals can be determined from incremental lines of dental cementum and their reproductive history obtained from examining their reproductive organs. Hence, from a random sample of females, one can establish the age distribution of the population, the age of first breeding, and the proportion of mature females that breed each year. From this, one can work out the size of the adult population from the observed number of pups.

Sampling

The need for sampling

It is sometimes possible to survey the whole of a population: one can count all the trees in a small woodland, estimate the entire population of fish in a small lake, or count all the calls made by frogs in an isolated marsh. More commonly, one can study only part of a larger area, subsequently generalising from the sample to the whole. For example: counting the trees in a few 1-ha blocks of forest, to estimate the average number of trees per hectare in the whole forest; estimating the number of fish present in a few pools in order to estimate the total number of fish in all the pools in a catchment; counting the number of whales migrating past an observation post on a few days in order to estimate the total number passing through the entire migration season.

For the inferences that one draws about the whole from one's samples to be valid, sampling must follow certain principles. For effort not to be wasted, efficient methods must be used. The main principles and methods are presented here. Cochran (1977) gives more detail.

For sampling to work, the samples must be representative of the whole. If they are not, then the generalisation to the whole will produce biased results; thus if one chooses to study a particularly damp part of a forest, it will have more mosquitoes in it than average, so an extrapolation from its mosquito population to that of the whole forest will exaggerate the estimated number for the whole forest. It is important to define both one's samples and one's total area of study carefully. Otherwise, one can neither ensure that the samples are representative nor safely generalise from the samples to the whole.

It is also important to recognise that sampling may take place at more than one level. In the fish-pond example above, the study pond is taken to be a sample of all 57 ponds in the catchment; but the estimate of the fish population in that pond may itself come from a mark–recapture study, itself based on several samples of fish from the pond. In this section on sampling, we treat all the work that produces a single population estimate (or index) for a particular time and place as a single sample, even if it is itself based on several samples at the lower level.

Replication

The need for replication

An investigator who proceeds to estimate the total number of trees in a forest from the number in one hectare, or who assumed that the total number of fish in a series of 25 similar ponds was just 25 times the number in one of the ponds, or who claimed that he knew the average number of helminth parasites in fish of a particular species because he had caught a single fish and counted its helminths would be laughed at. Why? Because we know that a single sample unit can only provide an imprecise estimate for the whole study population and, however carefully chosen, it may not be representative. To increase precision and representativeness, one needs more than one sampling unit.

Replication of sampling units has another essential function: it allows the precision of the generalisation to the whole study area to be measured. Thus one can not only estimate the total number of trees in the forest, the total number of fish in all 25 ponds, or the average helminth burden per fish, but one can place confidence limits on these estimates.

In several places above, methods have been described for estimating the precision of population estimates in individual units – for example, for an estimate of the number of fish in a pond based on mark–recapture methods. The precision of the estimate for the whole study area will be greater when the individual sample estimates are more precise. It is not, however, possible to assess overall precision from the precision of estimates for individual sample units. This is because overall precision depends not only on individual precision but also on the amount of variation between individual sample units. The latter can only be assessed if one has sampled more than one unit.

How many replicates?

Generally speaking, the precision of the overall estimate depends on the square root of the number of replicate samples. Thus, to halve the width of the confidence interval, one needs to quadruple the number of replicates. Considerations of cost will determine how many samples can be taken. These issues are explored in more detail for particular sampling designs below.

Table 2.6. *Two hundred and fifty random numbers*

Block no.	1	2	3	4	5
	41530	97900	58557	73058	28651
	06574	10732	62978	09675	33356
1	86198	54445	05595	08445	23506
	03381	62043	59755	55082	69309
	44452	98628	96650	70454	29352
	79317	34204	19936	01927	44427
	92933	55336	05370	54991	41203
2	18645	24577	87864	94257	30397
	48784	64416	71630	66256	17507
	91095	31323	45772	12202	84280

Ensuring that samples are representative

Random sampling

Suppose that one has decided to count the trees in ten 1-hectare blocks in a forest, in order to estimate the number of trees in a whole forest that extends to 50 000 hectares. How does one choose which ten of the 50 000 to sample, to ensure that the sample is representative? Surprisingly, one does *not* choose ten 1-hectare blocks that appear to be 'typical'. Experience shows that such judgemental samples rarely are truly typical. Furthermore, it is not possible to estimate the precision of the overall estimate properly from judgemental samples. Nor does one use 'haphazard' approaches, such as stabbing at a map of the forest with one's eyes closed or, for a smaller-scale investigation, throwing a sample quadrat over one's shoulder; experience again shows that such methods do not provide random samples.

Randomness requires that each potential sample unit has an equal chance of being included in the sample. The best way to ensure this is to give each potential sample unit a number and then choose which numbers to include by using a table of random numbers (or a string of random numbers generated by a computer). The numbers in such tables are arranged such that every digit has the same 10% chance of occurring at every position in the table and so that there are no patterns in the numbers. Table 2.6 is a small table of random numbers; the larger ones provided in statistical textbooks usually have the numbers arranged in blocks, as here. Box 2.20 shows, through simple examples, how to use such a table to choose sample units.

Box 2.20. **Using random-number tables to choose which units to sample**

Choosing a starting point in the table

If one always starts to use the table at the same point (such as the top left corner), one will get the same random numbers in all of one's studies. This may be undesirable. It is better to choose a different starting point each time. In general, haphazard methods, such as stabbing at the table with one's eyes shut, after a colleague has rotated the table, are satisfactory. A better method is to use a die, as follows: roll the die four times, noting the numbers obtained; if you roll a six, ignore it and roll again; use the first number to define which main block (on the horizontal axis) to use, the second to define which column in that block to use, the third to define which block (vertically) to use, and the fourth to define which row in that block. Thus the rolling sequence 4, 4, 1, 2 would result in starting at 7 if one were using Table 2.5 (contained in the sequence 09675 in the second row). This starting point is used in the examples below.

Choosing three out of ten sample units

Label the sample units 0–9. The first three digits in the table of random numbers, reading from the starting point (7, 5, 3), identify which three units to choose.

Note that if one had required a sample of four, the first four random digits would have been 7, 5, 3, 3. Since (for the methods presented here) the same sample unit cannot be included twice, one ignores the second 3 and reads the next digit; in Table 2.5 it happens also to be a 3; one reads the next digit – a 5 – which has also occurred already; the next digit is new (6), giving the fourth sample number at last.

Choosing out of up to 100 sample units

Label the units 0–99. Proceed as before but now read the digits in pairs: 75, 33, 35 are the first three from the chosen starting point in Table 2.5. At the end of a line, simply continue onto the next line uninterrupted: thus the fourth number in our sequence would be 68. Numbers such as 08 are read as 8. Numbers beyond the required range are ignored.

Choosing out of more than 100 sample units

If there are 101–1000 units, one generates 3-digit numbers in the same way as above, and so on for larger numbers of units. If one arrives at the bottom of the table, one can simply continue from the last digit in it to the first in the top left corner. It is not a good idea, however, to continue so far that one starts using the same numbers again; it is better to use a larger table of random numbers.

Box 2.21. Choosing sample units from a two-dimensional distribution

Imaginary example: choosing one-hectare sample areas from a woodland. The shaded area in the diagram below represents the woodland and the lines of the grid that is superimposed on it are 100 m apart.

Number the grid rows and columns.

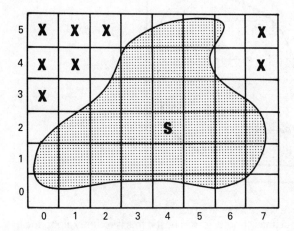

Get a pair of numbers from a random-number table in the usual way. The first defines the column in the grid, the second defines the row. If the numbers were 4, 2, this would define the sample square as the one marked S in the example.

The process is continued until the required number of units has been defined.

If grid squares that do not contain part of the study area (marked X in the diagram) are chosen by this process, they are ignored.

Note that it is unnecessary to mark out the grid on the ground; as long as one corner is identified on the ground, and the compass orientation of the grid is fixed, then it is possible to measure out the locations of the chosen sample units.

Sample units are often naturally arranged in two dimensions. Box 2.21 shows a simple way of choosing random units in this case. A similar technique may be applied if the sample units are based on points rather than areas (in studies using trapping webs or point transects, for example): the pairs of numbers derived from the table are then used to define the distances to the sample points from an arbitrary base point. Only points falling in the study area are utilised.

To extend the method to line transects is also straightforward (Box 2.22).

Box 2.22. **Randomising line transects**

Imaginary example: the shaded area represents a study area; the box is a superimposed rectangular area. The dots represent the starting points of transects, with the transect lines extending rightwards from them.

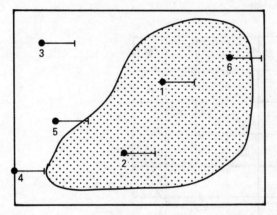

Decide in what direction transect lines should lie; it is best if this is roughly along the long axis of the study area. (Note that it is not necessary to randomise transect directions.)

Choose random points within the superimposed rectangular area. These are the starting points of the transects. Points such as 1 and 2 in the example provide straightforward transects. Points 3 and 4 provide transects lying wholly outside the study area; they should not be included in the sample. (Note the importance of allowing a complete transect length between the leftmost part of the study area and the left boundary of the rectangle, in the case of transects that run rightwards from the chosen points.)

For points such as 5 and 6, only those parts of the transects that lie within the study area should be surveyed.

Some methods that might appear at first sight to provide random samples do not do so: one needs to be careful. For example, laying down one or more lines at random across an intertidal area and including in the study all those individuals of a bivalve species that happen to be crossed by the line will bias the sampling towards the larger individuals. So will taking into the study those individual trees in a forest whose canopies cover any one of a set of random points. Such size-biased sampling is best avoided; while the resultant data can be used, the methods of interpreting them are beyond the scope of this book.

Regular sampling

It is often convenient to take samples at regular intervals – on the same day of the week, at fixed distances along a river, or at all the intersections of a rectangular grid. This results in two problems. First, for statistical reasons, it may produce confidence limits for the overall population estimate that are unreliable. Second, and more obviously, it will produce biased results if the regularity of the sampling coincides with a natural regularity in the distribution of the organisms. For example, waterbird numbers using a lake may be less than average on Sundays because hunting activity is most intensive on Saturdays; if counts are always made on Sundays, they will underestimate the average number present over the whole week. (Such data can be used to generalise only about Sundays, not all days of the week.) There may be similar, and often more subtle, regularities in spatial distribution. Regular sampling should be avoided if possible.

People often believe that regular sampling is advantageous because it distributes the samples all over the study area, thus ensuring that, in aggregate, they are truly representative. Since such distribution can be achieved by stratified sampling (see below), regular sampling is not required.

Regular sampling is justified in one situation: when one is trying to map variation in abundance of an organism across a study area as well as to estimate its total abundance. The advantages for mapping of a regular distribution of the sample sites may then outweigh its disadvantages for population estimation.

The problem of inaccessible areas

Parts of a study area may be inaccessible. If so, they must effectively be excluded from the study. Having made a proper estimate of numbers in the accessible part of the area, one may then make a less formal estimate of the numbers in the accessible area, based on whatever knowledge is available. An estimate of the total population is then possible but the fact that only part of the population has been properly studied. Thus a (stratified) random sample of Britain outside large urban areas, where the survey methods could not be applied, gave an estimate of the total number of social groups of badgers *Meles meles* in non-urban Britain as $41\,894 \pm 4398$ (95% confidence limits); from other knowledge of badger populations in large urban areas (thought to total several hundred), the total number of social groups in Britain may be estimated as around 42 thousand.

It is also not uncommon that part of the study area is more difficult or more costly to survey than the rest. Stratified random sampling is the solution (see below).

Unavoidable deviations from randomness

Suppose that one is sampling invertebrate animals on a mudflat. In taking a sample from a point, one may considerably disturb the immediately surrounding area. Thus the area within a certain distance of each sample is one within which one should not take another sample. Unless this results in a substantial proportion (more than 10%) of the study area being excluded from the sampling, this causes no problems. Otherwise, an adjustment has to be made in the calculation of confidence limits of the resultant population estimates.

Large-scale projects, such as national surveys that depend on volunteers being prepared to go out to survey sample localities, rarely result in all of the chosen localities being surveyed. Volunteers may, for example, not be interested in surveying certain habitats or places where they are likely to find few interesting animals or plants. This could cause serious bias. Even if the reason for failing to cover some areas is that landowners refuse access, bias can result, for access may be more likely to be refused for certain types of land. All surveys where inclusion of individual randomly chosen samples is optional are liable to bias.

The example of a parasitologist surveying the helminths in fish presents another class of problem. The fish that are caught for study are not, in the formal sense, a random sample of the fish population. Indeed, if helminth burdens alter catchability, they may be a biased sample. They can be regarded as a representative sample of the more easily catchable individuals in the population. Since it is impossible to take formally random samples from most natural populations, one has to make do with such catchable samples and use one's knowledge of natural history to judge how well they represent the populations from which they are drawn. If one catches animals using two different methods and the results differ, this is evidence that at least one of the methods is biased, though it does not indicate which; similarity of the results does not indicate that bias is absent, since there may be similar bias in the two methods.

The size and number of sampling units

Uniform size is desirable

The 'size' of a sample unit may be determined by its area or volume but there are other measures of size, such as the length of a line transect, the amount of time spent making counts, the number of traps in a grid, or the number of points in a set of point transects or point counts. One has to accept variation in size if sampling units are naturally defined (such as the fish in ponds or the ponds themselves) or if it is convenient to use existing features to define sampling units rather than an artificial grid (such as using fields rather than superimposed grid squares in an agricultural area). Where, on the contrary, the investigator can control the size of the sample units, it is best to choose units of uniform size. This generally maximises the precision obtained for a given effort.

If sample units are of different sizes, one needs to consider whether unweighted or weighted means are the more informative for one's purpose. A weighted mean allows for the difference in size of the sample units. For example, if ponds with areas 6, 7 and 8 hectares contain 366, 469 and 568 fish (i.e. 61, 67 and 71 fish/ha), then the weighted mean density of fish is

$$(61 \times 6 + 67 \times 7 + 71 \times 8)/(6 + 7 + 8) = 66.8 \text{ fish/ha}$$

whereas the unweighted mean is simply

$$(61 + 67 + 71)/3 = 66.3 \text{ fish/ha}$$

In the rest of this chapter we assume that sample sizes are equal, so weighted means can be ignored.

The effective area sampled at each point in point-to-object methodology varies according to local density. If density varies simply because of the random distribution of the population, the resultant over- and under-representation of the various densities cancels out. Otherwise it does not. This is why point-to-object methods produce biased estimates (unless the population is randomly distributed).

What size is best?

If sample units are too small, the number of organisms in them may be systematically biased by an 'edge effect'. We have seen how this can happen in trap-based methods, as a result of immigrant animals being trapped, but it can happen in many other ways: for example, observers often tend to include individuals that lie on (or even just outside) the sample boundary. However, if sample units are so large that each one requires a huge effort, it will not be possible to take many samples during the whole study, thereby reducing the precision of the overall estimate of average numbers. The balance to be struck between a few large units and many small ones will vary from case to case. It should be considered at the planning stage of any study. The discussion on cluster sampling (p. 87) is relevant to such considerations.

Note that the edge effect can be minimised by using square sample areas rather than rectangular ones, since squares have smaller ratios of edge to area. Practical considerations may dictate other shapes, however. For example, if one has to descend a cliff on ropes in order to count the plants growing on it, the easiest technique will probably be to count the plants in a strip transect on each side of the rope, from the top to the bottom of the cliff.

How many units to sample

Box 2.23 shows how to calculate both the sample size that one needs to take in order to obtain a PRP of the required magnitude (or less) and also the PRP attainable for a predetermined sample size.

Note that making the prediction requires some knowledge of the standard deviation and mean of the number of animals or plants per sample unit, or at least of their ratio. (The latter is often fairly stable across populations, so one can use values obtained in other studies; count data typically have values in the range 0.5–1.0, so values in this range can be used if one has no better information available). Two methods of estimating the required sample size are shown in Box 2.24. The first requires some estimate of the $s/\bar{N}$ ratio (or a guess at its value); the second requires a preliminary survey to estimate s and $\bar{N}$, followed by a top-up survey to take the total sample size to the required number (the final estimate of $\bar{N}$ and its confidence limits being based on both parts of the survey combined).

Box 2.23. **Estimating the sample size needed for a random sampling to attain a fixed PRP**

Basics

Q = the required percentage relative precision (Box 2.1)

$\bar{N}$ = mean number of organisms per sample unit

s = standard deviation of number of organisms per sample unit

m' = sample size required for there to be a 95% chance of obtaining a PRP of Q or less

Method 1: $s/\bar{N}$ estimated or guessed in advance

Calculate a first approximation to m':

$$m_0 = \left(\frac{200}{Q}\right)^2 \left(\frac{s}{\bar{N}}\right)^2$$

If $m_0 < 25$, $m' = m_0 + 2$; if $50 > m_0 > 25$, $m' = m_0 + 1$; if $m_0 > 50$, $m' = m_0$.

Imaginary example: estimated $s/\bar{N} = 0.8$; $Q = 20\%$:

$$m_0 = \left(\frac{200}{20}\right)^2 (0.8)^2 = 64; \; m' = 64$$

To have a 95% chance of achieving a PRP of 20%, one should plan to take a sample of 64.

Method 2: requiring a preliminary survey

m_1 = number of sample units in the preliminary survey

$\hat{N}_1$ = mean estimated from this preliminary sample

s_1 = standard deviation estimated from this preliminary sample

m^+ = additional number of sample units required to have a 95% chance of obtaining a PRP of Q or less (i.e. $m' = m_1 + m^+$)

$$m^+ = \left(\frac{200}{Q}\right)^2 \left(\frac{s_1}{\hat{N}_1}\right)^2 \left(1 + \frac{2}{m_1}\right)$$

Imaginary example:

$m_1 = 10$

$\hat{N}_1 = 3.2$

$s_1 = 4.0$

$$m^+ = \left(\frac{200}{25}\right)^2 \left(\frac{4.0}{3.2}\right)^2 \left(1+\frac{2}{10}\right) = 120$$

Thus, following a preliminary sample of 10, a further 120 sample units are needed, if the required precision is to be obtained with a probability of 95%.

Note: if $(m_1 + m^+)$ is less than 50, it should be adjusted like m_0 (see Method 1, above).

Adjustment for large sampling fraction

M = total number of sample units available in the study area

When the required sampling fraction (m'/M) is large, the required sample size can be reduced to

$$m^{\bullet} = m'\left(\frac{M}{M+m'}\right)$$

This adjustment is worth making if (m'/M) is greater than 0.1 – i.e. if the number of units needed is more than 10% of the total number available in the whole study area.

Precision attained for a fixed cost

C_T = the fixed total cost
c = cost of taking one sample
Q' = precision (PRP: Box 2.1) attained (with 95% probability) for a fixed total cost

Method 1 ($s/\bar{N}$ estimated or guessed in advance):

$$Q' = (200s/\bar{N})\sqrt{c/C_T}$$

Imaginary example: $C_T = £10\,000$, $c = £5$, prior estimate of $s/\bar{N} = 2.0$:

$$Q' = (200 \times 2.0)\sqrt{5/10\,000} = 9\%$$

Method 2 requiring a preliminary survey):

$$Q' = (200s_1/\hat{N}_1)\sqrt{c(m_1+2)/m_1(C_T - cm_1)}$$

Imaginary example: $C_T = £10\,000$, $c = £5$, $m_1 = 10$, $s_1 = 4.0$, $\hat{N}_1 = 3.2$:

$$Q' = (200 \times 4.0/3.2)\sqrt{5(10+2)/10(10\,000 - 5 \times 10)} = 6\%$$

Box 2.24. Means and confidence limits from a set of samples

Basics

m = number of sample units studied

M = total number of potential sampling units in the study area (may be effectively infinite – see text)

$\hat{N}_i$ = estimated number of organisms in the ith sample

Note: the same method can be applied to indices or densities.

Examples

(1) Number of Grain Aphids *Sitobion avenae* on 50 random shoots of Winter Wheat *Triticum aestivum* in an English field in May 1988. (Data from N. Carter, personal communication.)

Data (f = number of shoots with $\hat{N}$ aphids):

$\hat{N}$	0	1	2	3	4	5	6	7	8	9	10	11	12	13	14	15	16	21	26	33
f	14	4	12	3	3	2	1	0	3	0	1	0	2	0	1	0	1	1	1	1

(in this table, $f=0$ for all missing values of $\hat{N}$)

Preliminary calculations:

																				Sums
$f\hat{N}$	0	4	24	9	12	10	6	0	24	0	10	0	24	0	14	0	16	21	26	33	233
$f\hat{N}^2$	0	4	48	27	48	50	36	0	192	0	100	0	288	0	196	0	256	441	676	1089	3451

(2) Number of newly emergent shoots of Bramble *Rubus fruticosus* in ten random 1.28 × 1.28 m quadrats within a 9 × 9 m study plot in a British woodland. (Hutchings 1978; Diggle 1983.)

											Sums
$\hat{N}$	0	1	4	4	7	7	7	8	11	13	62
$\hat{N}^2$	0	1	16	16	49	49	49	64	121	169	534

Calculation of mean and variance

$$\bar{N} = \sum \hat{N}_i / m$$
$$= 233/50 = 4.66 \text{ (example 1)}$$
$$= 62/10 = 6.20 \text{ (example 2)}$$

$$s^2 = \frac{\sum(\hat{N}_i - \bar{N})^2}{m-1} = \frac{\sum N_i^2 - (\sum N_i)^2/m}{m-1}$$
$$= (3451 - 233^2/50)/49 = 48.27 \text{ (example 1)}$$
$$= (534 - 62^2/10)/9 \quad = 16.62 \text{ (example 2)}$$

Calculation of standard error when *M* is effectively infinite

Proposed as in example 1, where there was a huge (and uncounted) number of shoots.

$$s_{\bar{N}} = \sqrt{\frac{s^2}{m}}$$
$$= \sqrt{48.27/50} \qquad = 0.98 \text{ (example 1)}$$

Calculation of standard error when *M* is finite

Proposed as in example 2, where there were only 49 sample units in the study area.

$$s_{\bar{N}} = \sqrt{\frac{s^2}{m}\left(1 - \frac{m}{M}\right)}$$
$$= \sqrt{\frac{16.62}{10}\left(1 - \frac{10}{49}\right)} \qquad = 1.15 \text{ (example 2)}$$

Calculation of confidence limits for the estimated mean

$$95\% \text{ confidence limits} = \bar{N} \pm t \times s_{\bar{N}}$$

where *t* is Student's *t* for 5% significance and $m-1$ degrees of freedom.

Example 1: limits $= 4.66 \pm (2.01 \times 0.98) = 2.69$ and 6.63

Example 2: limits $= 6.20 \pm (2.26 \times 1.15) = 3.60$ and 8.80

Estimation of means and total population sizes

Means and confidence intervals

The calculation of the mean and variance of numbers per sampling unit (or of an index or of density) is straightforward (Box 2.23). To calculate the standard error and confidence limits of the mean one needs to consider the relationship between the samples and the whole study area from which they have been drawn. It is generally easy to consider the area as comprising *M* potential sampling units, of which *m* are actually sampled: the imaginary example in Box 2.21 is a clear-cut case. In other situations there is a potentially infinite number of units to be sampled: the example in Box 2.22 is such a case, for the number of potential starting points that could be distributed in the study area is limited only by the precision with which their positions are measured out. When there is a finite number (M) of potential sampling units, then the sampling fraction (m/M) needs to be taken into account in calculating the standard error of the mean (Box 2.24).

Total population size

Provided that the data come from a random sample, the means and confidence limits calculated from the sample units can be generalised to the whole study area. Hence, if $\bar{N}$ is the mean number of animals or plants per sample unit, the best estimate of the total size of the population in the whole study area is

$$\hat{N}_T = M\bar{N}$$

Confidence limits of this estimate are similarly obtained, by multiplying the limits for $\bar{N}$ by M.

When distributions are not normal

The normal distribution is an abstraction on which many statistical procedures are based. Unfortunately, the frequency distributions of counts, population indices and densities are rarely normal. Figure 2.12 shows the data for the first example in Box 2.23, which are clearly not normal. Such departure from normality has no effect on the estimation of the means and variances but it does affect the standard error and confidence limits, which will therefore be misleading if one applies standard methods to non-normal data.

A common solution to the problem of normality is to transform the data. For counts, using square roots of the data rather than the data themselves often produces a normal distribution. This allows one validly to carry out statistical analyses such as *t*-tests and analyses of variance, which depend on the observations being normally distributed. It is commonly assumed that one can calculate confidence limits for the mean of the untransformed data by back-transforming the limits calculated for the transformed data: if one has used the square-root transformation, one squares the limits of the transformed data. Unfortunately, such back-transformed limits may be unreliable and should never be regarded as other than approximate.

Another problem with using transformations of counts is that count data often include a particularly large number of zero values (e.g. Figure 2.12). No transformation can normalise such a distribution. There are four ways of dealing with such problems. One is to increase the size of the sample units, if possible. This will often reduce the number of zero counts. Another solution, which has the same effect, is to combine related sampling units; for example, if the several sample counts from each of the farms included in the survey that produced the data in Figure 2.13a are combined, the mean counts per farm are substantially closer to a normal distribution, with far fewer zero values (Figure 2.13b).

It may be possible to fit a statistical distribution other than the normal to the data; the Poisson and negative binomial distributions are often applied to count data. If so, one can

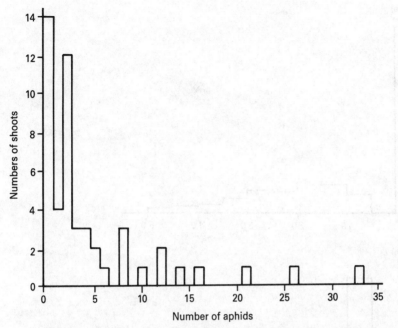

Figure 2.12 Number of wheat shoots (in a sample of 50) with various numbers of aphids *Sitobion avenae* in an English field in May. (Data from N. Carter, personal communication.)

calculate confidence limits appropriate to that distribution. Finally, one can obtain confidence limits by using computer-intensive 'randomisation' methods, such as the jack-knife or bootstrapping (Manly 1991). These methods are beyond the scope of this book.

Cluster sampling

What is cluster sampling?

Suppose one is counting ferns within 10×10-m sample quadrats in a forest measuring approximately 30×50 km. Because of the size of the forest, it takes a long time to reach each randomly chosen sample location. Once there, however, counting the ferns takes just a short time. Would it not be more efficient to take a number of samples at each point – perhaps counting all the 10×10-m quadrats in a 40×40-m grid (Figure 2.14a)? This is cluster sampling – taking a set of samples at each of a number of random positions. Other examples would be taking mud samples at the corners of a square quadrat laid down in random positions on a shore (Figure 2.14b) or placing standard sets (e.g. a 2×5 array at 1-m spacing) of pitfall traps in random locations. Clustering may be temporal rather than spatial: for example, obtaining an average index of frog numbers for a month by counting the number of frog calls in four 15-minute periods on each of four evenings (the evenings being randomly positioned during a month) rather than by counting the calls in one period on each of 16 random evenings.

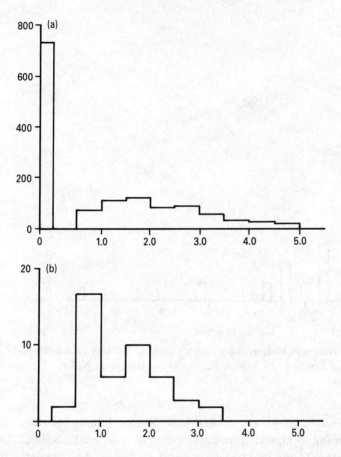

Figure 2.13 Frequency distributions of number of pairs of Chaffinches *Fringilla coelebs* per km of hedgerow on 46 British farms: (a) individual values for 1331 100-m sample lengths of lengths of hedgerow; (b) mean values for each farm. (Data from British Trust for Ornithology, personal communication.)

The problem with cluster sampling is that the sample units within each cluster are not independent of each other: the numbers of ferns in adjacent 10 × 10-m quadrats are likely to be more similar than those in quadrats that are further apart; the numbers of frog calls in 15-minute periods on the same evening are likely to be more similar than in 15-minute periods on different evenings. As a result, if one were to use the standard methods of assessing the precision of means, which assume that the samples are all independent, one could be seriously misled. The only solution to this problem is to combine the data for all the samples in the cluster and then treat the cluster totals as single sample measurements, using the usual statistical methods. Thus one's sample size is the number of clusters, not the total number of individual sample units.

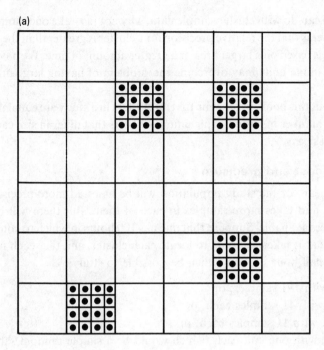

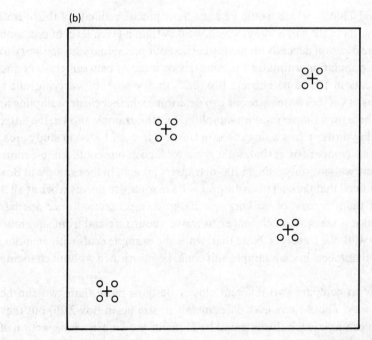

Figure 2.14   Two forms of cluster sampling: (a) cluster areas (the large squares) are chosen at random and all sample units (small squares) in each are sampled; (b) points are chosen at random (+) and samples (○) taken in a fixed pattern relative to each.

If this is the best one can do with cluster sample data, why not just take one sample unit in each cluster? The answer is that the relative precision for a cluster is greater than the precision of a single unit, since it is based on a larger area or a greater amount of time. We have already seen that combining sample units may overcome the problem of having large numbers of zeros in the data set.

Note that, throughout this book, we assume that all clusters in a study are equal in size (i.e. they contain the same number of basic sample units). Clusters that differ in size can be used but data analysis is less easy.

Considerations of cost and precision

The overall total or mean for the study population will be assessed more precisely if one samples more clusters and takes more samples in each of them. But there will be a limit imposed by the resources available. Suppose that one has 100 hours in which to complete the fieldwork in a study, that it takes $4\frac{1}{2}$ hours to locate each cluster, and that each individual sample takes another half hour. It would then be possible to study:

> 1 cluster, with 191 samples; or
> 4 clusters, with 41 samples each; or
> 10 clusters, with 11 samples each; or
> 20 clusters, with one unit each (which would be a simple random sample).

Which would be best? That is, which would give the most precise estimate of the overall mean? (Or, which amounts to the same thing, which would attain a given level of precision for the lowest cost?) The decision depends on having carried out preliminary survey work to determine variation in population estimates (or densities or indices) between clusters and between the sample units in the same cluster. Box 2.25 shows how, by carrying out a pre-survey based on clusters of the chosen size one can determine whether cluster sampling at that scale is better or worse than simple random sampling. For the example shown, the latter would consist of counting birds on just a single 200-m transect in each 1 × 1-km study area. Note that the basis of this comparison is that, for a given total cost, one could sample more 1 × 1-km areas if one were walking only a single 200-m transect in each. In the example in Box 2.25, it is arbitrarily assumed that the cost of visiting a 1 × 1-km square to do any work at all is about 50 times greater than the cost of walking one 200-m transect section once one has located the square – since it takes so much longer to travel about England from square to square than actually to walk the transects. Note that, while the example deals with an index, estimates of density or of numbers in each sample unit could be substituted without changing the methods.

Box 2.26 shows how to compare two different cluster designs; more than two can be compared in the same way. The clusters may differ merely in size (as in Box 2.26) but they may differ in layout: for example, each cluster could be 25 plants growing in a long section of a single row of a crop or 30 plants growing in short sections of three adjacent rows.

Box 2.25. **Choosing between cluster sampling and simple random sampling**

Basics

Treat each cluster as a 'group' or 'treatment' in a standard analysis of variance, with the individual sample units providing the data in the form of estimates of population sizes for each unit. Carry out the analysis of variance in the usual way, to estimate

s_b^2 = mean square between clusters
s_w^2 = mean square within clusters (='error' mean square)
$F = s_b^2/s_w^2$ (as usual)

Obtain an estimate of the costs of the fieldwork:

c_M = basic cost of sampling a cluster, ignoring the cost of taking each individual
     sample
c_u = additional cost per sample unit

Thus, if there are U sample units per cluster, the total cost of sampling each cluster is

$c_M + Uc_u$

Decision criterion

Cluster sampling is superior if

$$\left(\frac{c_M}{c_u}+U\right)\Big/\left(\frac{c_M}{c_u}+1\right)<1+(U-1)F$$

Example

Counts of Skylarks *Alauda arvensis* seen on 200-m sections of transects on 16 1 × 1-km areas of English farmland in spring 1994, there being 10 transects systematically placed on each area: the 1 × 1-km areas are the clusters and the 200-m sections are the sampling units. (Data from British Trust for Ornithology, personal communication.)

Summarised data:

Cluster (i)	1	2	3	4	5	6	7	8	9	10	11	12	13	14	15	16
$\sum I_i$	16	4	14	2	2	16	15	23	20	10	14	15	24	8	13	15
$\sum I_i^2$	44	4	40	2	2	32	33	59	70	24	32	39	76	22	45	57

$\sum I_i$ = sum, for the *i*th cluster, of the counts
$\sum I_i^2$ = sum, for the *i*th cluster, of the squared counts

For the analysis of variance we require

$$\sum(\sum l)^2 = (16^2 + 4^2 + \ldots + 15^2) \quad = 3441$$
$$(\sum\sum l)^2 = (16 + 4 + \ldots + 15)^2 \quad = 44\,521$$
$$\sum\sum l^2 = 44 + 4 + \ldots 57 \quad\quad = 581$$

If there are m clusters, sums of squares are

$$SS \text{ (between cluster)} = \sum(\sum l)^2/U - (\sum\sum l)^2/mU$$
$$= 3441/10 - 44\,521/160 \quad = 65.84$$
$$SS \text{ (within cluster)} = \sum\sum l^2 - \sum(\sum l)^2/U$$
$$= 581 - 3441/10 \quad\quad = 236.90$$

Mean squares and their ratios are

$$MS \text{ (between cluster)} = SS(\text{between})/(U-1)$$
$$= 65.84/9 \quad\quad = 7.32$$
$$MS \text{ (within cluster)} = MS(\text{within})/U(m-1)$$
$$= 236.90/150 \quad\quad = 1.58$$
$$F = 7.32/1.58 \quad\quad = 4.63$$

In this example, an approximate value of $\dfrac{c_M}{c_u}$ is 50. Hence

$$\left(\frac{c_M}{c_u} + U\right) \Big/ \left(\frac{c_M}{c_u} + 1\right) \quad = (50+10)/51 \quad = 1.2$$
$$1 + (U-1)F \quad\quad = 1 + (9)4.63 \quad = 37.8$$

Thus cluster sampling, following the decision criterion above, is superior to simple random sampling.

Box 2.26. **Choosing between different cluster designs**

Basic

U_1, U_2 = the number of sampling units per cluster, in each of the two designs
s_1^2, s_2^2 = the variances of the cluster totals (or means), in each of the two designs
c_1, c_2 = the costs of sampling the whole cluster, in each of the two designs

Decision criterion

The design that gives the greatest precision for a fixed total cost, or the least total cost for a fixed precision, is the one with the lowest value of

$$cs^2/U^2$$

For a comparison of just two designs, the ratio of the values of this expression is the ratio of costs (for the same standard error of the overall mean) and also the ratio of standard errors (for the same cost).

Example

The Skylark data in Box 2.25. We compare the ten-segment clusters of Box 2.25 with an alternative of five-segment clusters (i.e. five 200-m transect segments per 1×1-km square). The 16 cluster totals ($\sum l_i$ in Box 2.25) were

10-segment   16, 4, 14, 2, 2, 16, 15, 23, 20, 10, 14, 15, 24, 8, 13, 15
5-segment   11, 3, 9, 2, 1, 6, 10, 12, 6, 3, 8, 5, 10, 4, 5, 8

$$s_1^2 = [(16^2 + 4^2 + \ldots + 15^2) - (16 + 4 + \ldots + 15)^2/16]/15 = 43.90$$
$$s_2^2 = [(11^2 + 3^2 + \ldots + 8^2) - (11 + 3 + \ldots + 8)^2/16]/15 \quad = 11.46$$

If c_M is the basic cost of visiting a 1×1-km square and c_u the additional cost of each transect segment, then

$$c_1 = (c_M + u_1 c_u) = \left(\frac{c_M}{c_u} + u_1\right)c_u$$

$$c_2 = (c_M + u_2 c_u) = \left(\frac{c_M}{c_u} + u_2\right)c_u$$

In this example, we take c_M/c_u as 50, so

$$c_1 = 60c_u; \quad c_2 = 55c_u$$

Thus

$$c_1 s_1^2/U_1^2 = (60c_u \times 43.90)/100 = 26.3c_u$$
$$c_2 s_2^2/U_2^2 = (55c_u \times 11.46)/25 \quad = 25.2c_u$$

Sampling five 200-m segments in each 1×1-km square is thus just superior to sampling ten, though the costs (for a fixed precision) are only 4% less.

Further analysis of the data on which the examples in Boxes 2.25 and 2.26 are based shows that the costs (for a given precision) of sampling based on clusters of various sizes, as a percentage of that for cluster sizes 6 and 7 (which are the best sizes), are:

cluster size	1	2	3	4	5	6	7	8	9	10
relative cost	330	214	160	135	107	100	100	107	108	111

Thus, in the range 5–10, it makes little difference exactly which cluster size is used. This is a typical result. The exact magnitude of the difference between the basic cost of a cluster and the additional cost of each sample unit is also rather unimportant if the difference is large. This means that it is safe to apply conclusions about optimal cluster size from work in one area or at one time to similar places or times, since it is unlikely that the differences between places or times will have much effect on the optimum.

How many clusters to sample

One works out how many clusters one needs to sample by applying the methods of Box 2.23, treating clusters as individual samples (since this is how one treats them to work out overall means and confidence limits for the whole study area).

Multi-level sampling

What is multi-level sampling?

Because the data on individual sample units within a cluster are combined, information on the variance between sample units is lost in cluster sampling; the precision of one's estimates is thereby reduced. One can overcome this by sampling only some of the units within each cluster – and by doing so at random. Thus, one might randomly locate 40×40-m grids in a forest and take a random sample of some of the sixteen 10×10-m quadrats within each of them (Figure 2.15).

We shall refer here to the larger sample units (e.g. the 40×40-m grids) as major units and the units within them (e.g. the 10×10-m quadrats) as minor units. Our treatment will be restricted to two levels of sampling, though the principles can be extended to as many levels as one wishes: for example, 40×40-m grids may be located within random 1×1-km squares in a forest and one might then take 5-cm-diameter soil cores randomly within 10×10-m quadrats within the 40×40-m grids if one were studying the soil microflora – an example of four-level sampling. The number of levels depends on practical convenience.

Further levels may be added subsequent to the initial sampling, by subsampling. Thus it would rarely be convenient to take the material from the entire 5-cm soil core for cultivation of microflora in the lab. One would take random subsamples, which add another level.

Both the costs and precision of multi-level sampling tend to be intermediate between simple random sampling and cluster sampling. It is often superior to both.

Optimal distribution of sampling

The precision of the overall mean depends on both the variance between the mean values for major units and the variance between minor units within major units (the second variance being assumed to be constant over all major units); it depends also on the number of units sampled at each level. Given a fixed limit to the cost of the work (or a certain level of precision required), there is clearly an optimal distribution of sampling effort between major and minor units, which depends on both the variances and the costs of sampling at the two levels. To determine the optimal number of minor units to sample within each major, as shown in

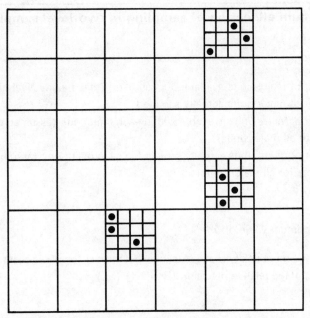

Figure 2.15 Two-stage sampling: major units (large squares) are randomly chosen and minor units (small squares) sampled at random within each.

Box 2.27, one requires some knowledge at least of the ratio of the two variances and of the ratio of the costs at the two levels. These may be obtained through a preliminary survey or through general knowledge of the situation one is studying. As with cluster sampling, the difference in realised precision between the absolute optimal sampling pattern and a pattern close to it is often slight, so precise knowledge of the variances and costs is not required.

Once the optimal number of minor units per major has been determined, one can work out the corresponding number of major units, depending on whether one is working to a fixed cost or to a required level of precision (Box 2.27). Note that, for the latter, one requires estimates of the two variances individually, not just their ratio.

Estimates from two-level sampling

Box 2.28 shows how to estimate the mean numbers of organisms per sample, with confidence limits, for two-level sampling. It assumes that all of the minor units are of the same size, that each of the major units contains the same number of minor units, and that the same number of minor units is sampled from each major unit. All these assumptions can be relaxed but the calculations become less straightforward and it is generally best to have such even distribution of sampling effort. Note that the number of minor units within major units can be effectively infinite, if the samples are point counts, for example, since there is an infinite number of points within an area.

Box 2.27. **The optimum allocation of sampling in two-level sampling**

Basics

M = number of major units available for sampling (may be effectively infinite)

m = number of major units actually sampled

U = number of minor units available within each major unit (assumed to be the same for all major units)

u = number of minor units actually sampled in each major unit (assumed to be the same for all major units)

We need to calculate

u_{opt} = the optimum value of u

From this, we can estimate the value of m needed to carry out the task for a fixed cost (m_c) or to achieve a required percentage relative precision (Box 2.1) (m_p).

Example

An ecologist wished to estimate the density of worms of a particular species on a shore extending over $150\,000\,m^2$. In a preliminary survey, he established five randomly placed 2×2-m quadrats and took four random 20×20-cm cores from each. Thus the sampling structure was

$M = 150\,000/(2 \times 2) = 37\,500$

$m = 5$

$U = (2 \times 2)/(0.2 \times 0.2) = 100$

$u = 4$

The means and variances of the numbers of worms per core within the quadrats were

quadrat	mean	variance
1	5.00	8.667
2	5.50	3.000
3	7.75	6.917
4	2.50	5.667
5	8.50	19.000

These can be used to calculate (in the usual way) three further quantities:

$\bar{\bar{N}}$ = mean of the means   = 5.850 worms/core

s^2_M = variance of the means = 5.675

s^2_M = mean of the variances = 8.650

The investigator also found that, on average, it took half an hour to locate each quadrat and 6 minutes to take each core and count the worms. These can be used as measures of costs:

$$c_M = 0.5 \text{ hours} = \text{basic cost per major unit sampled}$$
$$c_U = 0.1 \text{ hours} = \text{additional cost for each minor unit sampled}$$

Estimation of u_{opt}

$$u_{opt} = \sqrt{\frac{c_m}{c_u} \bigg/ \left(\frac{s_M^2}{s_U^2} - \frac{1}{U}\right)}$$
$$= \sqrt{\frac{0.5}{0.1} \bigg/ \left(\frac{5.675}{8.650} - \frac{1}{100}\right)} = 2.78 = 3$$

Thus three cores should be taken in each quadrat.

Note that if either

$$u_{opt} > U$$

or if

$$s_M^2 s_U^2 < \frac{1}{U}$$

then one should sample all of the minor units within each major unit (thus simplifying the design to simple random sampling of major units).

Sample size for a fixed cost

Suppose that the limit of expenditure on the survey is to be C. Then:

$$m_c = C/(c_m + c_u U_{opt})$$

If $C = 8$ hours in our example,

$$m_c = 8/(0.5 + 0.1 \times 3) = 10$$

Thus, the optimum allocation of sampling effort (to achieve the most precise estimate of the mean population density) in this case, if 8 hours are available, is to use 10 quadrats and take 3 cores in each.

Sample size to achieve a required percentage relative precision (Q)

$$m_P = \left[s_M^2 + s_U^2\left(\frac{1}{u} - \frac{1}{U}\right)\right] \bigg/ \left[\frac{s_M^2}{M} + \left(\frac{\bar{\bar{N}}}{200}Q\right)^2\right]$$

If the PRP in our example is 20%,

$$m_P = \left[5.675 + 8.650\left(\frac{1}{3} - \frac{1}{100}\right)\right] \bigg/ \left[\frac{5.675}{37500} + \left(\frac{5.850}{200}20\right)^2\right]$$

$$= 24.7 = 25$$

Thus the optimal allocation of sampling effort in this case, to achieve a PRP of 20%, is to use 25 quadrats and take 3 cores in each.

Box 2.28. **Calculating overall means, totals and confidence limits from two-level sampling**

Example

As in Box 2.26. Suppose that the ecologist carried out a main survey using 10 quadrats, with 3 cores in each (to achieve the most precise estimates within a total cost of 8 hours – see Box 2.26). Thus

$$M = 37\,500$$
$$m = 10$$
$$U = 100$$
$$u = 3$$

Suppose that the various means and variances (see Box 2.26) were

$$\bar{\bar{N}} = 5.240 \text{ worms/core}$$
$$s^2_M = 5.392$$
$$s^2_U = 10.017$$

Estimation of overall mean and total numbers

The best estimate of the overall mean number of animals per minor unit is simply $\bar{\bar{N}}$.

The best estimate of the total number in the entire study area is $\bar{\bar{N}}MU$.

In the example,

$$\bar{\bar{N}}MU = 5.240 \times 37\,500 \times 100 = 19.65 \text{ million worms}$$

Confidence limits

The standard error or $\bar{\bar{N}}$ is

$$s_{\bar{\bar{N}}} = \sqrt{\left(1 - \frac{m}{M}\right)s^2_M\bigg/m + \left(1 - \frac{u}{U}\right)s^2_U\bigg/um}$$

In this example,

$$s_{\bar{\bar{N}}} = \sqrt{\left(1 - \frac{10}{37\,500}\right)5.392\big/10 + \left(1 - \frac{3}{100}\right)10.017\big/30}$$
$$= 0.9289$$

Approximate 95% confidence limits ae

$$\bar{\bar{N}} \pm t \times s_{\bar{\bar{N}}}$$

using Student's *t* for 5% significance and for $m(u-1)$ degree of freedom.

In this example, the limits are

$$5.240 \pm 2.086 \times 0.9289 = 3.30 \text{ and } 7.18 \text{ worms/core}$$

The standard error and confidence limits for the total number are obtained by multiplying the values of $\bar{\bar{N}}$ by *MU*.

In this example, the confidence limits for the total are thus 12.4 and 26.9 million worms on the whole beach.

Box 2.28 is presented in terms of numbers, including estimation of total numbers in the whole study area. As usual, indices and densities can be treated in the same way.

Stratified sampling

What is stratified sampling?

It is often obvious even before a study takes place that there are systematic variations in population density across the study area. If so, then it is usually valuable to divide the study into subareas that differ in density and to sample randomly within each. Such subareas are strata. They are not the same as the major units of two-level sampling, which are random subdivisions, not subdivisions systematically chosen to reflect the variation within the study population.

The strata do not need to be individually continuous. One might, for example, divide a study area into high- and low-density areas that happened to be intermingled (Figure 2.16).

Stratification allows separate estimates of the means and variances to be made for each stratum. Its main value, however, is that it allows the overall mean to be estimated with much greater precision. For example, the confidence limits for the example used in Box 2.29 are only half as wide as if the same number of samples had been taken without stratification. Using simple random sampling, such an improvement in sample size would have required a fourfold increase in sample size. Stratification is also valuable if the costs of sampling are different in different parts of the study area; places further from the base of operations may, for example, entail greater travel time and costs. Investigations in which different organisations are responsible for carrying different parts of the study area are also well treated as stratified studies – each organisation's area being a stratum.

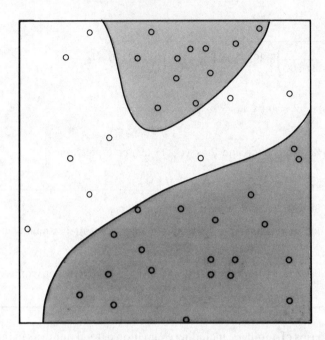

Figure 2.16   Stratified sampling: the study area is divided into strata that differ in mean density of the organism being investigated and samples taken randomly within each. In this example there are two strata, one of high density (shaded), which in this case happens to encompass two separate areas, and one of low density; the investigator has chosen to sample more intensively in high-density stratum.

How to stratify

If administrative organisation imposes the stratification, one accepts the stratification that it presents. Otherwise one chooses to subdivide the study area in the way that minimises the within-stratum variance in population density (and, correspondingly, maximises the differences between strata). How can one do this ahead of the fieldwork? One way is to base it on a rough preliminary survey. For example, one may be able to estimate numbers roughly by simple observation of what animals or plants can be seen, without precise counts; or one may get a rough idea of densities using simple point counts, in order to stratify an area prior to using more precise methods such as mark–recapture; or one might even simply divide the area into places apparently with and without the species. Alternatively, one can stratify according to habitat or ecological characteristics with which the organism's density is likely to be correlated. These methods will not provide such effective stratification as if one were able to use precise knowledge of the variation in population density but they will provide much better results than simple random sampling. So long as there is some difference between the stratum means, however small, stratification is advantageous.

Stratification may lose its advantages if the number of strata is large relative to the total number of sampling units (so that each stratum contains just a few units). It is usually sufficient to use only 3–6 strata.

If the costs of sampling are different in different parts of the study area, such cost differences can be used to define strata, especially if the cost differences are clear. For example, if studying an area including both mainland and islands, the latter may be much the more costly to sample because of the need to use boats to reach them; stratifying into mainland and

---

Box 2.29. **Optimal allocation of sampling effort for stratified sampling**

Basics

For each stratum we require:

M_h = total number of potential sample units in stratum h
m_h = number of units actually sampled in stratum h
$W_h = M_h/\sum M_h$ = stratum weight for stratum h
$\bar{N}_h$ = estimated mean number of organisms per sampling unit in stratum h
s_h = standard deviation of number of organisms per sampling unit in stratum h
c_h = cost of taking one sample in stratum h

Example

Badgers *Meles meles* surveyed in Great Britain by counting the number of main setts in 1 × 1-km squares. The country was divided into 8 land-class groups, defined by environmental characteristics. Urban areas were omitted from the survey (Reason *et al.* 1993). Values of c_h used here are arbitrary. Individual values of m_h actually used are provided here for completeness – they are required for the estimations (Box 2.30). Values of $\bar{N}_h$ used here are estimates from the main survey: for a planning exercise, one would have to guess at their values.

Land-class group	M_h	m_h	W_h	$\bar{N}_h$	s_h	c_h
I	44 762	627	0.1979	0.472	0.9014	4
II	28 513	352	0.1261	0.174	0.5066	4
III	39 004	425	0.1724	0.172	0.4329	4
IV	27 180	308	0.1202	0.205	0.5089	4
V	25 775	224	0.1140	0.031	0.1947	4
VI	23 814	143	0.1053	0.014	0.1673	9
VII	31 550	313	0.1395	0.073	0.3008	4
VIII	5 573	63	0.0246	0.016	0.1269	9
Sums	226 171	2455	1.0000			

The optimum relative effort in each stratum

The proportion of the sampling effort that should optimally be made in the hth stratum is given by

$$\frac{m_{h\,(opt)}}{\sum m_h} = \frac{\bar{N}s_h/\sqrt{c_h}}{\sum(\bar{N}_h s_h/\sqrt{c_h})}$$

Calculation of the relevant values for the badger example the second column is used later):

Land-class group	$\bar{N}_h s_h$	$\dfrac{\bar{N}_h s_h}{\sqrt{c_h}}$	$\bar{N}_h s_h \sqrt{c_h}$	$\dfrac{m_{h\,(opt)}}{\sum m_h}$
I	0.425 48	0.212 74	0.850 96	0.5853
II	0.088 14	0.044 07	0.176 28	0.1212
III	0.074 46	0.037 23	0.148 92	0.1024
IV	0.104 33	0.052 16	0.208 66	0.1435
V	0.006 04	0.003 02	0.012 08	0.0083
VI	0.002 34	0.000 78	0.007 02	0.0049
VII	0.021 96	0.010 98	0.043 92	0.0302
VIII	0.002 03	0.000 68	0.006 09	0.0042
		0.361 66	1.453 93	1.0000

Thus, for optimal effect, 59% of the samples should have been taken in land-class group I (a large stratum with a high mean and variance) but only 0.4% from group VIII (a small stratum with a small mean and variance).

Note that, whatever the $m_{h\,(opt)}$ value, at least two samples should be taken from each stratum, otherwise the variance cannot be calculated.

Sample sizes for a fixed cost

Suppose that the study must be contained within a fixed cost C and that the basic cost of mounting the study, before any samples at all are taken, is c_0. The requisite total number of samples (provided they are allocated to strata in the optimum proportions shown above) is then

$$\sum m = (C - c_0) \sum(\bar{N}_h s_h/\sqrt{c_h})/\sum(\bar{N}_h s_h\sqrt{c_h})$$

Using an arbitrary value of $C - c_0 = 10\,000$ and the sums from the table above,

$$\sum m = 10\,000(0.361\,66/1.453\,93) = 2487$$

The proportions $(m_{h\,opt}/\sum m_h)$ are multiplied by this value to obtain the optimum number of samples to take in each stratum, for an overall cost C.

Sample size for a required precision

Further sums are required, as follows:

Land-class group	$W_h s_h$	$A = \dfrac{W_h s_h}{\sqrt{c_h}}$	$B = W_h s_h \sqrt{c_h}$	$C = W_h s_h^2$
I	0.178 40	0.089 20	0.356 79	0.160 81
II	0.063 88	0.031 94	0.127 75	0.032 36
III	0.074 63	0.037 32	0.149 26	0.032 31
IV	0.061 17	0.030 59	0.122 34	0.031 13
V	0.022 19	0.011 10	0.044 39	0.004 32
VI	0.017 62	0.005 87	0.052 86	0.002 95
VII	0.041 97	0.020 98	0.083 93	0.012 62
VIII	0.003 12	0.001 04	0.009 36	0.000 40
Sum		0.228 04	0.946 68	0.276 90

If V is the required value of $s_{\bar{N}}^2$, then

$$\sum m = = \frac{\sum A \sum B}{V + \sum C / \sum M_h}$$

If working in terms of percentage relative precision, substitute V, as usual, by $(\bar{\bar{N}} Q/200)^2$, where $Q =$ required PRP and $\bar{\bar{N}}$ is the overall mean.

Thus for a PRP of 5% in the badger survey, using a guess of $\bar{\bar{N}} = 0.2$,

$$\sum m = (0.946\,68 \times 0.228\,04)/[(0.2 \times 5/200)^2 + (0.276\,90/226\,171)] = 8232$$

---

island areas is likely to be useful. If there is variation in both density and cost, both need to be considered but variation in density is generally the more important.

Optimal allocation of sampling effort

How many samples should one take from each stratum? If the costs of sampling and within-stratum variance in density are the same in each stratum, then sampling with the same intensity in each stratum is best – that is, it gives the greatest precision for the least cost. If the variances are unknown (and unguessable) in advance, one should assume that they are all the same. If the variances are known to differ, however, one should sample more intensively in the strata that are the more variable. If costs differ, one should sample more where the costs per sample are lower. Box 2.29 shows exactly how the effort should be distributed.

In practice, the exact distribution of sampling effort makes little difference to the precision attained. This means that one can use a rough idea of the variances in each stratum to guide the allocation of sampling effort.

Estimation of overall mean and totals

To estimate the overall mean, one needs to estimate the mean and variance (in the usual way) for each stratum separately. One also needs to calculate the stratum weight (the proportion of the total study area that the stratum comprises) and the stratum sampling intensity (the proportion of the potential sample units in each stratum that were actually sampled). Box 2.30 shows how these values can be used to obtain estimates of mean density and of the total population of the study area, with confidence limits.

Box 2.30. **Estimation of overall mean and confidence limits from a stratified random sample**

Estimate of overall mean number per sample unit

Using the same symbols and example as Box 2.28, the estimate of the overall mean is

$$\bar{\bar{N}} = \sum W_h \bar{N}_h$$
$$= 0.1979 \times 0.472 \ldots + 0.0246 \times 0.016 = 0.185\,21 \text{ setts/km}^2$$

Standard error and confidence limits

Further calculation is required for each stratum, in particular:

$$s_h' = W_h^2 s_h^2 (1 - f_h)/m_h = \text{contribution of the } h\text{th stratum to the overall standard error}$$
$$g_h = M_h (M_h - m_h)/m_h$$

These are tabulated, together with a number of derived values:

Land-class group	$s_h'^*$	g_h^{**}	$g_h s_h^{2**}$	$\dfrac{(g_h s_h^2)^2}{(m_h - 1)}^{***}$
I	5.0052	3.1508	2.5603	10.4718
II	1.1443	2.2811	0.5853	0.9761
III	1.2864	3.5405	0.6635	1.0383
IV	1.2005	2.3713	0.6142	1.2287
V	0.2178	2.9401	0.1114	0.0557
VI	0.2158	3.9420	0.1104	0.0858
VII	0.5571	3.1486	0.2850	0.2602
VIII	0.0153	4.8742	0.0785	0.0993
Sum	9.6424		5.0086	14.2159

Note that the values in this table have been adjusted by orders of magnitude for convenience of layout. Hence:

* values in this column should be divided by 10^5,
** values in these columns should be multiplied by 10^6,
*** values in this column should be multiplied by 10^9.

The standard error of the overall mean is

$$s_{\bar{N}} = \sqrt{\sum s'_h} = \sqrt{9.6424 \times 10^{-5}} = 9.820 \times 10^{-3}$$

95% confidence limits are calculated in the usual way, as:

$$\bar{\bar{N}} \pm t s_{\bar{\bar{N}}}$$

where t is Student's t for the 5% significance level and d degrees of freedom, the value of d being

$$d = (\sum g_h s_h^2)^2 / \sum[(g_h s_h^2)^2/(m_h-1)]$$
$$= (5.086 \times 10^6)^2/(14.2159 \times 10^9) = 1766$$

Thus, for this example, the limits are

$$0.18523 \pm (1.98 \times 9.820 \times 10^{-3}) = 0.18523 \pm 0.01944$$
$$= 0.16578 \text{ and } 0.20467 \text{ setts/km}^2$$

Estimate of the total population size

As usual, simple multiplication by the total size of the study area provides the estimate of the size of the total population and its confidence limits:

$$\hat{N}_T = 226171(0.18521 \pm 0.01944)$$
$$= 41889 \pm 4398 \text{ setts in non-urban Britain}$$

Strip transects within or across strata

As noted above, it may be more convenient to have long, narrow quadrats than square ones. For example, it may be quicker to walk 400 m in a line in a forest counting all the ferns within 2 m of one's path (total 1600 m²) than to find a random 40 × 40-m quadrat and count the ferns within that. Such line transects may run across areas within which there are considerable variations in numbers of the organism being studied. That variation may be clinal, with a gradient in numbers from one edge of the study area to the other. How should the line transect be placed in relation to the cline?

Figure 2.17a shows one answer to this question: to divide the cline up into strata, the divisions lying parallel to the contours of population density, and to take randomly

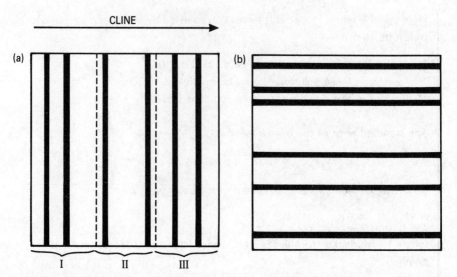

Figure 2.17   Alternative methods of aligning strip transects in relation to a cline in population density: (a) random positions along strata (I–III) placed across the cline; (b) random positions parallel to the cline.

positioned line transects within each stratum, their long axes also lying parallel to the contours of density. Figure 2.17b presents an alternative: do not stratify but simply take randomly positioned line transects along the direction of the cline (cutting across its strata).

Which arrangement is superior? If one wishes to obtain a description of the cline, then the second method is better. But if one wishes to get an overall estimate of average density or numbers, the answer is less clear-cut. If one has sufficient prior knowledge to be able to stratify effectively, then the two methods are about equally efficient statistically; the choice should thus be determined largely by practical convenience. If, however, the strata do not fit the variations in density closely then the first method is not as good as the second.

Surveillance

Surveillance means observing whether, and to what extent, something changes over time. There is no space here to consider all aspects of surveillance but some broad principles must be considered.

If one can census the whole study area, surveillance merely entails repeated censuses at an interval appropriate for the study in question. If one's study is sample-based, one may simply take independent random samples at each time. Box 2.31 shows how to calculate the change in population size (or density or index) between two times, with confidence limits. (The same method can be used for comparing two study areas.)

Particularly if sampling intensity is great, the same sample unit may, by chance, be chosen for inclusion in the sample on more than one occasion. If sampling disturbs the study

---

Box 2.31. **Estimating change in numbers between two sampling occasions, based on independent samples**

Basics

$\bar{N}_1$, $\bar{N}_2$ = estimated mean numbers (or densities or indices) per sample unit on the two occasions

s_1^2, s_2^2 = estimate variances of numbers per sample unit on the two occasions

m_1, m_2 = sample sizes on the two occasions

Example

Chaffinches *Fringella coelebs* on 65 English farms in 1992 and 1993. The data were number of breeding territories per hectare, measured on a random sample of 20 farms in each year, the samples in the two years being independent, with five of the farms being included in both samples by chance (British Trust for Ornithology data).

$$\bar{N}_1 = 0.329 \text{ territories/ha, } s_1^2 = 0.029\,66$$
$$\bar{N}_2 = 0.272 \text{ territories/ha, } s_2^2 = 0.029\,03$$

Estimation of the change and its confidence limits

The change in mean numbers is simply estimated:

$$\bar{d} = \bar{N}_2 - \bar{N}_1 = 0.272 - 0.329 = -0.057 \text{ territories/ha}$$

(the minus sign indicates a decrease in numbers)
The standard error of this difference is

$$s_{\bar{d}} = \sqrt{\frac{(m_1 - 1)s_1^2 + (m_2 - 1)s_2^2}{(m_1 + m_2 - 2)} \left(\frac{m_1 + m_2}{m_1 m_2}\right)}$$
$$= \sqrt{\frac{19 \times 0.029\,66 + 19 \times 0.029\,03}{(20 + 20 - 2)} \left(\frac{20 + 20}{20 \times 20}\right)}$$
$$= 0.0542$$

95% confidence limits are

$$\text{Difference} \pm t \times s_{\bar{d}}$$

where t is Student's t for 5% significance and $(m_1 + m_2 - 2)$ degrees of freedom.

In this example, the limits are

$$-0.057 \pm 2.024 \times 0.0542 = -0.167 \text{ and } +0.053$$

(i.e. we are 95% confident that the average change over all 65 farms lay between a decline of 0.167 territories/hectare and an increase of 0.053 territories/hectare).

population so much that it has not returned to its natural state by the time the second sample is taken, the site cannot be included in the second sample. Should this happen more than occasionally, adjustments need to be made to the confidence limits of the estimate.

If it is possible to sample the same site repeatedly, it is usually much more effective to use exactly the same sample units on each occasion than to engage in independent sampling on each occasion. Box 2.32 shows how to calculate the difference in population between two occasions. It uses the same example as in Box 2.31. The great improvement in precision obtained by using the same sites rather than independent samples is substantial – and quite typical.

Box 2.32. **Comparison of estimated population densities at two different times, based on constant sample locations**

Basics

M	= total number of sampling units in the study area
m	= number of units actually sampled
$\hat{N}_{i1}$, $\hat{N}_{i2}$	= estimated numbers (or densities or indices) on the two occasions in the ith sample unit
d_i	= $\hat{N}_{i2} - \hat{N}_{i1}$

Example

As in Box 2.31, but the same random sample of farms was used in both years. The densities on individual farms, and their differences, are tabulated:

farm number i	densities 1992	1993	difference d_i	d_i^2
1	0.188	0.200	−0.012	0.000 144
2	0.000	0.000	0.000	0.000 000
.	.	.	.	
.	.	.	.	
.	.	.	.	
20	0.798	0.681	0.117	0.013 689
Sums			0.210	0.060 651

Estimation of the change and its confidence limits

The differences are treated as data items for the usual calculations of the mean and its confidence limits.

The change in mean numbers is simply estimated:

$$\bar{d} = \sum d_i / m = 0.210/20 = 0.0105 \text{ territories/ha}$$

The variance between sample units in the change in numbers is

$$s_d^2 = \left[\sum d_i^2 - \frac{(\sum d_i)^2}{m}\right]\Big/(m-1)$$

$$= \left(0.060\,651 - \frac{0.21^2}{20}\right)\Big/19 = 0.003\,07$$

The standard error and 95% confidence limits of $\bar{d}$ are

$$s_{\bar{d}} = \sqrt{\frac{s_d^2}{m}\left(1-\frac{m}{M}\right)} = \sqrt{\frac{0.003\,07}{20}\left(1-\frac{20}{65}\right)} = 0.010\,309$$

$$CL = \bar{d} \pm t \times s_{\bar{d}} = 0.0105 \pm 2.093 \times 0.010\,309 = -0.0111 \text{ and } +0.0321$$

(t is Student's t for 5% significance and $m-1$ degrees of freedom)

Notes

1. The estimate of the difference obtained here is different from that in Box 2.31, simply because they are based on different samples.
2. The confidence limits here are much narrower than those in Box 2.31, because of the same farms being sampled in the two years.

In respect of sampling systems (random, stratified, etc.), the same principles apply to surveillance as to population estimation generally. Note, however, that the objective of surveillance is not to estimate numbers but to estimate changes in numbers. This may influence the design of one's survey. For example, it may be possible to divide the range of a species into two strata – a core area, in which it occurs extensively, and a peripheral area, in which its occurrence is sporadic. If merely estimating numbers at one time, one would do best to sample more intensively in the core area than in the peripheral area. But in many species, variations in numbers from year to year may be greater in the peripheral area than in the core of the range; if so, a progamme of surveillance would be more effective if sampling intensity were concentrated in the peripheral area.

Acknowledgements

I thank: Drs Julian Greenwood, Jim Fowler and Will Peach for checking and reviewing the chapter; Drs Nick Carter, Bob James, Keith Sutherland, Mike Fraser, Professor Les Underhill and the British Trust for Ornithology (especially Dr Dan Chamberlain and John Marchant) for providing data; Professor Stephen Harris for advising on the badger data; Drs Rob Elton and Derek Yalden for providing literature; Julie Sheldrake for word-processing; and Diane Alden for drawing the figures.

References

Anderson, D. R., Burnham, K. P., White, G. C. & Otis, D. L. (1983). Density estimation of small mammal populations using a trapping web and distance sampling methods. *Ecology* **64**, 674–580.

110   *Basic techniques*

Bibby, C. J., Phillips, B. N. & Seddon, A. J. E. (1985). Birds of restocked conifer plantations in Wales. *Journal of Applied Ecology* **22**, 619–633.

Buckland, S. T., Anderson, D. R., Burnham, K. P. & Laake, J. L. (1993). *Distance Sampling. Estimating Abundance of Biological Populations*. Chapman & Hall, London.

Burnham, K. P. & Overton, W. S. (1978). Estimation of the size of a closed population when capture probabilities vary among animals. *Biometrika* **65**, 625–633.

Burnham, K. P. & Overton, W. S. (1979). Robust estimation of population size when capture probabilities vary amongst animals. *Ecology* **60**, 927–936.

Caughley, G. (1977). *Analysis of Vertebrate Populations*. John Wiley & Sons, London.

Cochran, W. G. (1977). *Sampling Techniques*. 3rd edn. John Wiley & Sons, New York.

Craig, G. C. (1953). On the utilisation of marked specimens in estimating populations of insects. *Biometrika* **40**, 170–176.

Diggle, P. J. (1983). *Statistical Analysis of Spatial Point Patterns*. Academic Press, London.

du Feu, C., Hounsome, M. & Spence, I. (1983). A single session mark/recapture method of population estimation. *Ringing & Migration* **4**, 211–226.

Edwards, W. R. & Eberhardt, L. L. (1967). Estimating Cottontail abundance from live-trapping data. *Journal of Wildlife Management* **31**, 87–96.

Hutchings, M. J. (1978). Standing crop and pattern in pure stands of *Mercurialis perennis* and *Rubus fruticosus* in mixed deciduous woodland. *Oikos* **31**, 351–357.

Jolly, G. M. (1965). Explicit estimates from capture–recapture data with both death and immigration – stochastic model. *Biometrika* **52**, 225–247.

Krebs, C. J. (1989). *Ecological Methodology*. Harper Collins, New York.

Leslie, P. H. & Davis, D. H. S. (1939). An attempt to determine the absolute numbers of rats on a given area. *Journal of Animal Ecology* **8**, 94–113.

Manly, B. F. J. (1991). *Randomization and Monte Carlo Methods in Biology*. Chapman & Hall, London.

McCaffery, K. R. (1976). Deer trail counts as an index to populations and habitat use. *Journal of Wildlife Management* **40**, 308–316.

Numata, M. (1961). Forest vegetaion in the vicinity of Chosi. Coastal flora and vegetation at Chosi, Chiba Prefecture IV. *Bulletin of the Chosi Marine Laboratory of Chiba University* **3**, 28–48 [in Japanese].

Otis, D. L., Burnham, K. P., White, G. C. & Anderson, D. R. (1978). Statistical inference from capture data on closed animal populations. *Wildlife Monographs* **62**, 1–135.

Pollock, K. H., Nichols, J. D., Brownie, C. & Hines J. E. (1990). Statistical inference for capture–recapture experiments. *Wildlife Monographs* **107**, 1–97.

Reason, P., Harris, S. & Cresswell, P. (1993). Estimating the impact of past persecution and habitat changes on the numbers of Badgers *Meles meles* in Britain. *Mammal Review* **23**, 1–15.

Seber, G. A. F. (1982). *The Estimation of Animal Abundance and Related Parameters*. Charles Griffin, London.

Southwood, T. R. E. (1978). *Ecological Methods*. 2nd edn. Chapman & Hall, London.

Turner, F. B. (1960). Size and dispersion of a Louisiana population of the Cricket Frog *Acris gryllus*. *Ecology* **41**, 258–268.

Underhill, L. G. & Fraser, M. W. (1989). Bayesian estimate of the number of Malachite Sunbirds feeding at an isolated and transient nectar resource. *Journal of Field Ornithology* **60**, 381–387.

Wileyto, E. P., Ewens, W. J. & Mullen, M. A. (1994). Markov-recapture population estimates: a tool for improving interpretation of trapping experiments. *Ecology* **75**, 1109–1117.

Wood, G. W. (1963). The capture–recapture technique as a means of estimating populations of climbing cutworms. *Canadian Journal of Zoology* **41**, 47–50.

Zippin, C. (1956). An evaluation of the removal method of estimating animal populations. *Biometrics* **12**, 163–169.

3 Plants

James Bullock

Furzebrook Research Station, NERC Institute of Terrestrial Ecology, Wareham,
Dorset BH20 5AS, United Kingdom

Table 3.1. *Methods described in this chapter and their applicability to different types of plant*

⋆ usually applicable, + often applicable, ? sometimes applicable, and no symbol indicates that the method is never applicable to that plant type. The page number for each method is given.

Method	Trees	Shrubs	Herbs and grasses	Bryophytes	Fungi and lichens	Algae	Seeds	Page no.
Total counts	+	+	?					113
Visual estimates	⋆	⋆	⋆	⋆	⋆	+		113
Frame quadrats	+	+	⋆	⋆	⋆	+		115
Transects	⋆	⋆	⋆	?	?			117
Point quadrats			⋆	⋆				119
Harvesting	?	?	⋆	+	?	+		121
Plotless sampling	⋆	+	?					123
Seed-bank soil cores							⋆	124
Seed traps							⋆	128
Marking and mapping	⋆	⋆	⋆	?	?	?		130
Vegetation mapping	⋆	⋆	⋆					132
Phytoplankton						⋆		135
Benthic algae						⋆		137

Introduction

Most plant communities consist of individual plants arranged on a surface (e.g. soil or rock). These plants are sessile; i.e. they 'sit still and wait to be counted', as J. L. Harper put it. This makes some surveying jobs simple. For instance, it is very easy to wander through the vegetation and make a species list. However, species and individuals within species often vary enormously in size (i.e. biomass, photosynthetic biomass, height, horizontal spread etc.) causing problems in selecting the best measure of species abundance. The standard measure of abundance of animals is density, a count of individuals in a unit area. This can be used for plants but it has two drawbacks. It may be hard to distinguish individuals of clonal plants where the genetic individual (genet) may consist of connected ramets (e.g. tillers or shoots), especially if the connections are buried rhizomes. It is sensible and usual in this case to estimate the density of ramets rather than genets and this has the added benefit that ramets will show less size variation. Variety in the size of plants (sometimes due to clonal growth) will mean that density measures lose a large amount of information about the community under study. For instance, there may be equal numbers of individuals of two species in your study area but the species with a larger average size will have a greater importance for the ecological processes in the area. Imagine comparing a herb and a tree species purely in terms of density. For these reasons measures have been devised which take into account both the size and density of plants.

Cover is a measure of the area covered by the above-ground parts of plants of a species when viewed from directly above. According to Grieg-Smith's (1983) commonly used description it is 'the proportion of ground occupied by a perpendicular projection onto it of the aerial parts of individuals of the species'. Because the vegetation may be layered the cover of all species often sums to more than 100%. Cover is a popular measure but as with other size-based measures it uses a particular definition of size, that of 'perpendicular projection'. Therefore, it favours species with spreading growth forms or larger leaves, and species which hold their leaves horizontally will have a higher cover than species with acute or obtuse leaf angles.

Biomass (usually the above-ground weight of the plants of a species) uses a more usual definition of size but also is biased, this time towards species with a greater tissue density or unit weight, such as woody species. The last common measure is frequency: the number of samples (usually frame quadrats, see p. 115) in which a species is found. This is extremely popular but is also a rather odd measure, being dependent on the cover of a species but also on the spatial pattern of the plants and on the size of the quadrats (see Frame quadrats, p. 115). Whichever measure is used, these descriptions and caveats should be kept in mind when you interpret your data.

The sessile nature of plants causes a clear and slow-changing spatial pattern in the distribution of species. Often patchiness in environmental variables, restricted dispersal of propagules and clonal growth all bring about a patchy distribution ('clumped', 'contagious' or 'overdispersed') of plants of a species. Your sampling strategy (see Chapter 2) must be designed to compensate for this patchiness and to give an accurate representation of the abundances of the species in the whole study area. If you keep this in mind, you will rarely have problems, although some measures (e.g. frequency and density by plotless sampling) will always be influenced by the form of the spatial distribution of species.

Species also show changes in relative abundance throughout the year particularly in response to seasons; what might be called temporal patchiness. This will be caused by temporal patterns in germination rates and/or in the growth form of plants. The latter is seen most spectacularly in plants which die back to, for instance, a bulb for part of the year. Species differ in their temporal behaviour. This should be accounted for in your sampling design or in your interpretation of the data.

Although plants are usually categorised into species, in surveys you could use a coarser system if you require a different type of information. This can be based on higher taxonomic groupings such as genera, families or even orders. Loose classifications based on systematics can be used: e.g. mosses, grasses, herbs, trees etc. Another classification method based on the growth forms of plants was developed by Raunkiaer (see pp. 1–4 of Kershaw & Looney 1983). Algae can be classified by their morphologies (e.g. single cell, colony or filament) and cell type rather than by individual species. For the methods described in this chapter these classification systems can be used instead of species.

These methods can also be used on some other organisms: fungi, lichens and sessile animals such as corals or encrusting bryozoans. They are possible in many benthic aquatic plant communities, although in deep water specialised equipment such SCUBA may be needed. Phytoplankton and the propagules of adult sessile plants do not sit still and wait to be counted and so different measures must be used for them.

Total counts

Assessing density of large or obvious plants that are at low density

Method

This technique is so simple it might be overlooked. Every individual of a species or a number of species in the study area is counted.

Advantages and disadvantages

Because the study area is usually several orders of magnitude larger than the plants, this technique is often much too time-consuming (imagine counting every plant in a 1-ha grassland). However, you should be able to use this technique if a species has a low enough density and is easily spotted (e.g. trees in a prairie) and the whole of the study area can be covered.

Biases

This measures the true density rather than sampling it and therefore has no biases.

Visual estimates of cover

Cover of species in any vegetation

Method

Visual estimates are made of the cover of the species either in the whole study area or in sample plots, such as frame quadrats (see p. 115). Different measures can be used. The

Table 3.2. *The Domin and Braun-Blanquet scales for visual estimates of cover*

Value	Braun-Blanquet	Domin
+	<1% cover	1 individual, with no measurable cover
1	1–5% cover	<4% cover with few individuals
2	6–25% cover	<4% cover with several individuals
3	26–50% cover	<4% cover with many individuals
4	51–75% cover	4–10% cover
5	76–100% cover	11–25% cover
6		26–33% cover
7		34–50% cover
8		51–75% cover
9		76–90% cover
10		91–100% cover

simplest is the classification: dominant, abundant, frequent, occasional or rare (DAFOR). These classes have no strict definition and you must decide on your own interpretation. Percentage cover can be estimated by eye either by creating your own percentage classes, e.g. in 10% or 25% steps, or by using those given in the Domin or Braun-Blanquet scales (Table 3.2). Remember that because vegetation is often layered, percentage cover values can sum to more than 100%. You may find it useful to divide the vegetation into layers, e.g. a bryophyte layer, a herb layer and a shrub layer, and make cover estimates separately for each layer.

Advantages and disadvantages

Visual estimates of cover are made more easily where you can look down on the vegetation. Cover may be hard to estimate with any degree of accuracy in tall vegetation such as scrub or forest, although it is possible if you can look up at the canopy and estimate the cover of individual trees.

The great advantage of this technique compared to more complicated surveying is speed. However, it can be inaccurate because of the subjectivity of the estimates. This means that different people or one person at different times may make different estimates. DAFOR is vague, and if several people are surveying then the meaning of the classes should be carefully agreed before you start. You should also be careful that your own interpretation of the classes does not change as the survey progresses. These precautions are also necessary if you use estimates of percentage cover. Often it is much easier and quicker to make visual estimates using frame quadrats (see p. 115) rather than estimating from the whole study area. Cover estimates made in the small area of a quadrat will also be less prone to error. The information obtained is limited but may be all that is needed for a straightforward site description or for a preliminary site survey. Percentage cover estimates give more infomation than DAFOR. If percentage cover is estimated, the use of very small classes of, say, 5% steps may give the impression of increased information but you may be fooling yourself that you can distinguish between these classes. Because these estimates are all scores you can only use non-parametric statistics on your data.

Biases

It has been suggested that more conspicuous species, such as those in flower or those forming clumps of individuals, may be given undeserved high cover estimates. This should not be a problem if you are careful.

Frame quadrats

Cover, density, biomass or frequency of species in any vegetation

Method

Often these are simply called quadrats. They are used to define sample areas within the study area and are usually four strips of wood, metal or rigid plastic which are tied, glued, welded or bolted together to form a square. It can be helpful to use bolts so that the quadrat can be dismantled for storage or transport. For aquatic macrophytes a wood or plastic frame will float and can be used to sample floating or emergent vegetation on the water surface. For large quadrats, over $4\,m^2$, a frame will be unwieldy and as an alternative you could measure out the quadrat using tape measures, folding rulers or string. Corners are marked by posts, and it is important to keep a constant quadrat shape, for example by using a set-square to measure out right-angles. Although a square is often used, the quadrat shape is unimportant as long as you know its area. For certain purposes the quadrat can be divided into a grid of equal-sized squares using regularly spaced lengths of string or wire.

There is a technique to determine the appropriate size of quadrat to give a representative sample for a study area (see Greig-Smith 1983) but the theory behind it is confused and it has little practical use. Experience has shown that different vegetation types requires different quadrat sizes. Vegetation with smaller plants, greater plant density or greater species diversity should require smaller quadrats. The sizes most often used are: $0.01 - 0.25\,m^2$ in bryophyte, lichen and algal communities (for instance, on rocks or tree bark), $0.25 - 16\,m^2$ for grassland, tall herb, short shrub or aquatic macrophyte communities etc., $25 - 100\,m^2$ for tall shrub communities and $400 - 2500\,m^2$ for trees in woods and forests. Different quadrat sizes can be used to survey different vegetation types within a study area, such as the understorey layer and canopy layer in a forest.

Quadrats are placed in the study area according to your sampling design and different measures can be used to survey the vegetation.

1. Density is measured by counting the number of individuals of the study species within the quadrat. Many plants will lie on the edge of the quadrat, and for this and other measures you must decide which to classify as inside the quadrat. Often only the plants rooted within the quadrat are counted.
2. Visual estimates of cover can be made within each quadrat rather than for the whole study area.
3. Frequency is a simple measure calculated as the percentage of the quadrats you have placed in which the species was present. For example, if 3 quadrats contained the species and 20 were placed then the species had a frequency of 15%. The number of plants in the quadrats is ignored. The 'shoot frequency' is measured if you define as present plants with any part inside the quadrat. If only rooted plants are counted this gives the 'rooted frequency'. A different measure is 'local frequency', which can be derived if the quadrat is subdivided into a grid and the percentage of grid squares containing the species is calculated. Local frequency is often used at points on a transect or where only a few quadrats can be used.
4. Biomass of species can be measured by harvesting (see p. 121).

Advantages and disadvantages

Frame quadrats are very easy to use and can be used in a wide range of studies. It can be difficult and time-consuming to measure out very large quadrats.

The different measures have different advantages and disadvantages. Those associated with the theoretical bases of density, cover and biomass measures are given in the Introduction. Counting individuals for density can be very time-consuming and difficult, unless the plants have a low density or you use very small quadrats, and is usually only used in studies of single species. Cover measures suffer from the problems of visual estimates (see p. 113). If, as is common, species have a non-random distribution over the study area, then estimates of cover, density or biomass from a *single* quadrat will be changed by the size of the quadrat. This is because larger quadrats will even out the patchiness in the vegetation more than smaller ones. Single quadrats are never used, and carefully planned sampling (Chapter 2) will eliminate these size effects.

Frequency is a very quick and easy method to use but the estimate of frequency will always be influenced by quadrat size. This is because frequency is a qualitative measure (of presence or absence in a quadrat) which is used to calculate a quantitative percentage and the quadrat, rather than being used to select a sample area, is used as the dimensionless unit of measurement. Therefore, larger quadrats will usually be more likely to find the study species and will give higher frequency estimates than smaller quadrats (Figure 3.1). Patchiness in species distribution will reduce the likelihood of a randomly placed quadrat finding the species and will therefore also reduce the frequency estimate. For these reasons you should take great care when interpreting frequency measures, especially when comparing different study areas. Local frequency has the same problems.

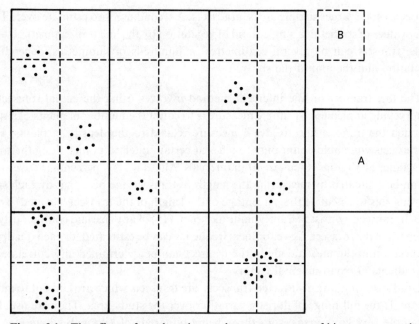

Figure 3.1 The effects of quadrat size on the measurements of biomass, cover, density and frequency. Quadrat A is four times the area of B. A single quadrat of size A will be more likely to hit an individual of a species (represented by a dot) than will a single B quadrat. If several quadrats are laid out (for simplicity the whole area is covered by quadrats in this example), the estimates of biomass, cover and density will be the same using A quadrats as if B quadrats are used. However, there will be more among-quadrat variation for the B quadrats. The different quadrat sizes give different estimates of frequency. This occurs for any distribution pattern of a species. In this example with a clumped distribution B quadrats give an estimate of 13/36 = 0.36 and A quadrats give 7/9 = 0.78.

Biases

Cover, density and biomass are discussed in the Introduction. Frequency can be biased against species with a more clumped distribution. Shoot frequency will be biased against smaller plants, but rooted frequency does not have this problem.

Transects

Variety of survey purposes in any vegetation

Method

Apart from the standard uses of transects (see Chapter 2) other transect-based methods can be used to survey vegetation. Transects are commonly used to survey changes in vegetation along an environmental gradient or through different habitats. This can be done using belt transects or, for larger sample areas, gradsects. A second use is to estimate overall density or cover values of species in a single stand of vegetation by the line transect method. The length of the transect can be several centimetres or hundreds of kilometres, depending on the vegetation and the aim of the study.

1. The line transect or line intercept method involves using the actual transect line as a surveying implement. A simple measure is to count the number of plants of a species that touch the transect line to give a measure related to the density of plants. For longer transects you could count only touches at certain interval points along the transect, for instance at 10-mm, 10-cm, or 10-m intervals. Alternatively, percentage cover (see p. 112) can be estimated by measuring the length of transect line occupied by each species and using this to calculate the percentage of the length of the transect 'covered' by a species.
2. Belt transects consist of frame quadrats (see p. 115) of any size laid contiguously along the length of the transect. Cover or local frequency can be estimated for each quadrat, and the variation in the measure along the transect can be determined and correlated with the gradients in environmental factors.
3. Gradsects, or gradient-directed transects, are transects which are laid out to intentionally sample the full range of floristic variation over the study area. They are usually used to sample very large areas, sometimes being hundreds of kilometres long (e.g. Austin & Heyligers 1989). To accomplish this the transect is usually positioned to lie along a steep environmental gradient, for example due to altitude, land use or geology.

Advantages and disadvantages

In certain vegetation types it may be easier to use the line transect method than frame or point quadrats. It can allow more productive sampling in sparse vegetation and can be more practical in tall vegetation. In either case, sampling will be speeded up. If the vegetation is at all dense then counting touches will take a very long time. If plants are tussocky, form definite clumps or are large and distinct then the length of transect occupied by a species can be measured reliably and simply. Cover estimates will be very difficult in vegetation where plants are small and intermingled, and different methods should be used. Counting touches by individual plants is not only difficult in dense and intermingled vegetation where it is hard to distinguish individuals but also produces a measure with ambiguous meaning. It resembles a cover measure, being determined by the density and size of a species (bigger plants are more likely to touch the transect), but cannot be expressed as percentage cover. It is hard to see any use in this measure. Although they produce very detailed data, belt transects can be very time-consuming, and it is worthwhile to consider whether more sparsely spaced quadrats will fulfil your needs. Gradsects, if used to assess the range of vegetation types over a large area, will yield more information than randomly placed transects and are commonly used to survey little-studied areas. However, a knowledge of the environmental gradients is needed, and you run the risk of being guided by environmental factors which are relatively unimportant in structuring the vegetation.

Biases

The count of touches or estimate of cover will often depend on the height of the line transect in the vegetation, different species having a different vertical and horizontal structure. Gradsects for an estimation of the range of floristic variation will be biased by the particular environmental factor used to describe the gradient.

Point quadrats

Estimating cover of grasses, herbs, mosses, etc. in short vegetation

Method

A point quadrat is a thin rod with a sharpened tip and should usually be made of metal for rigidity and strength. Good materials are thick gauge wire, welding rod, knitting needles or even bicycle spokes. The point quadrat is lowered vertically through the vegetation and different recording methods can be used to get different types of data. There are a number of measures but many are confusing and difficult to interpret and I shall not discuss them. The most popular and acceptable measure is of the percentage cover (see p. 112) of each species in the vegetation. To sample this you should identify the species of each living plant part that the tip (and only the tip) of the point quadrat hits on the way down to the soil surface. The data recorded from that point quadrat are only the presence or absence of each species, i.e. whether or not the point quadrat hits a species; the number of hits on a species is unimportant. If all the hits on a species were counted this would give a measure of the 'total cover' of a species, a measure which reflects the size of plants of a species as well as their abundance in the vegetation.

The theory behind the use of point quadrats is fairly simple. Frame quadrats (see p. 115) could be used to estimate cover by this method, i.e. the percentage of quadrats covered by the species. In this case the quadrat can be wholly covered by a species, partially covered or not covered at all. If the frame is large then many quadrats will be only partially covered, introducing an ambiguity to the measure of cover. As the frame size is decreased and becomes smaller relative to the size of the plants, the frames become less likely to be only partially covered and the cover estimate becomes more accurate. Point quadrats are theoretically a frame quadrat of infinitesimally small area, i.e. a point. A plant can only be present or absent in an area of zero diameter and therefore there are no cases of partial cover. This gives a true value for cover. Of course, point quadrats do not give absolute points but are cylinders with measurable diameters and you are actually sampling a small circle: the diameter of the point quadrat. This fact has caused considerable excitement among some ecologists. Greig-Smith (1983) and Kershaw & Looney (1983) discuss the theory behind point quadrats extensively. The estimate of percentage cover for different species can vary a lot with the diameter of point quadrat used (Goodall 1952). You could get a zero diameter point by using optical cross

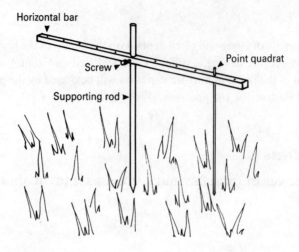

Horizontal bar

Point quadrat

Screw

Supporting rod

Figure 3.2 A point quadrat frame

wires, like those used in rifle sights, but these are well-nigh impossible to use in the field. The practical solution is to use point quadrats of as fine a diameter as possible. Steel wire of 1.5 – 2 mm diameter survives quite well in field use and a sharpened tip will narrow the point quadrat further. It is very important to use point quadrats of the same diameter in all your sampling, and if you want to compare your work with another study, find out what point quadrat diameter was used.

It is impossible to lower a point quadrat free-handed steadily and vertically through the vegetation while noting touches and identifying species. You should either stick the point quadrat into the soil and note touches along its length (it is important to have a very low diameter point quadrat for this) or use a 'point frame'. This consists of a sharpened metal supporting rod which supports a horizontal bar (Figure 3.2). The bar has holes along its length through which the point quadrat can be passed. The supporting rod is stuck vertically into the ground and readings are taken at the position of each hole in the bar. It is traditional, although not vital, for the frame to accommodate 10 point quadrats; the presence/absence readings from the 10 point quadrats are summed to give a score for the whole frame for each species (maximum = 10). You should never treat each of the 10 readings as independent measurements. Whatever method is used you should be careful not to disturb the vegetation when placing the point quadrat or taking readings; you could move plants onto or away from the point quadrat and cause errors in the sampling.

'Inclined point quadrats' are simply point quadrats which are lowered through the vegetation at an angle (usually 32.5°; see Warren-Wilson 1960) to the vertical. They are generally used in measurements of the canopy structure, which are particularly important in grazing studies. To measure canopy structure, not only are all the hits of the point quadrat noted, but also the height of each hit is recorded by marking a scale on the point quadrat.

Advantages and disadvantages

There is a sounder theoretical basis for using point quadrats to assess percentage cover than there is for visual estimates in frame quadrats, and canopy structure of short vegetation cannot be sampled any other way. Point quadrats are particularly useful in short vegetation, such as grasslands, and especially when it is difficult to distinguish individual plants. However, this technique can be very slow and fiddly especially in dense vegetation, and it involves crouching or lying on the ground for long periods. The vegetation should never overtop the point quadrat, and the point quadrats needed for some vegetation types such as hay meadows or tallgrass prairies will be too tall to be practical. Because a very small area is sampled very many samples may be needed to detect the rarest species.

Because leaves do not generally lie on the horizontal plane and inclined point quadrats enter the vegetation at an acute angle, this method will give more touches per point quadrat than will the vertical type. The inclined version will therefore give a more accurate sample of foliage area of each species, although this measure should not be confused with cover, which, being the proportion of ground covered by a species when viewed directly from above (see Introduction), can only be measured using vertical point quadrats. The measure of foliage area will also depend on the angle used. The practical problems of inclining the point quadrats at the same angle throughout the sampling programme and the ambiguity of the measure mean this approach is not usually worthwhile. Normal and inclined point quadrats can be used to measure the canopy structure, but because they give more touches, the latter seem to be preferred (e.g. Grant *et al.* 1985).

Biases

The biases involved in cover estimates are given in the Introduction.

Harvesting

Above-ground biomass of species in any vegetation

Method

The above-ground parts of the plants are cut at a certain height from the surface of the substratum, usually at or close to ground level. A knife, scissors, shears, saw or chainsaw may be used, depending on the vegetation type. Usually frame quadrats (see p. 115) should be used to define sample areas. The plant material should be taken to the laboratory, in bags or sacks when possible, and then sorted into species. Each species is either weighed as it is, giving a measure of 'fresh weight', or it is dried first, giving the 'dry weight'. The scales used will depend on the size of plants, but for any plants below the size of tall shrubs scales with an accuracy of at least 0.01 g should be used. Drying should be carried out in an oven at about 100°C for 1–2 days but if an oven is not available, natural drying for several days will work.

You must sort the species before drying or else the plants will be unidentifiable. It is important to remove or wash off any soil or detritus on the plant material before weighing.

Aquatic macrophytes may be harvested with a corer, which is effectively a quadrat frame that doubles as the harvesting tool. It is a sheet-metal cylinder with one rim sharpened and it is placed over the sample area and pushed into the substratum far enough to sever stems and roots. You could also use a grab to harvest an area of macrophytes.

Root harvesting (below-ground biomass) is too difficult and error-ridden to consider seriously.

Advantages and disadvantages

Harvesting has a large number of drawbacks and should be used only if you are certain that you need to measure the biomass of species. A particular instance of this is if vegetation is harvested to measure forage dry weight in grazing studies and you want to divide the biomass among the component species. This is known as 'destructive sampling', for obvious reasons, and should only be used in vegetation adapted to this sort of treatment (e.g. meadows), where the destruction is on such a small scale that it is unimportant, or where you and others do not mind the study area being destroyed. Large shrubs and trees are very difficult to harvest, transport and weigh. Very short vegetation, such as lawns, will be very difficult to harvest without massive error between samples caused by slight differences in cutting height. This method is really only appropriate for taller vegetation such as short shrub, aquatic macrophyte or meadow communities.

Cutting height is extremely important and you should usually cut at or close to ground level. However, the ground surface is rarely level and you may easily cut too high in some spots and too low (i.e. below the soil surface) in others, causing errors and contaminating the sample with soil. Corers for aquatic macrophytes cut below the substrate surface but there will still be variation in the depth of cutting. Grabs cut at very variable heights and also bring up a great deal of detritus and do not produce a reliable estimate of biomass at all.

Sorting of species can be very time-consuming and difficult, especially if the plants fall apart, leaving you with fragments to identify. Fresh weight is not a good measure. It varies with the moisture content of the plants and the moisture loss from cut plants means that fresh weight will be strongly affected by the time since harvest. Dry weight is a much better measure.

Biases

Low-lying species can easily be under-represented or even missed altogether if the cutting height is too high. If some species fall apart it is almost inevitable that you will lose and misidentify some plant parts. Fresh weight will bias the measure in favour of species with more effective moisture retention.

Plotless sampling

Estimating tree density in forests or woods

Method

A number of plotless sampling, or point-to-object, methods are possible. These are discussed more generally in Chapter 2, but I shall discuss only the two used commonly on trees. For either method a number of sample points are located at random in the study area. You can estimate the density either of all trees regardless of species or separately for each species. For the latter you can either lay out a different set of sample points for each species or use the same sample points and take separate distance measures for each species. The latter may be quicker. The minimum acceptable number of sample points will depend on the variation in the data, but as a rule of thumb you should use at least 50.

1. The nearest-individual method. The nearest tree to the sample point is located and the distance between it and the sample point is measured. The mean of the distances over all the samples is D_1 and the density of trees is worked out by the equation

$$\text{density} = 1/(2D_1)^2$$

2. The point-centred quarter method. Two perpendicular straight lines which cross each other on the sample point are measured out. This creates four quadrants; in each quadrant measure the distance to the nearest tree, as for the nearest-individual method. The orientation of the lines should be fixed in advance. A useful method is to use compass points or, if a transect line is used to locate the sample points, this can be used as the first line. The distances of these four trees are averaged and the mean of these averages over all the samples is D_2.

$$\text{density} = 1/(D_2)^2$$

Both methods involve the assumption that the distance is measured to the centre of the tree. You must estimate this distance. You must also decide whether or not to include saplings as trees in the survey. This is not usually done and you must decide at what size a sapling becomes a tree. This could be when it becomes part of the canopy layer.

Advantages and disadvantages

Neither of these methods gives a random sample of individual trees. Even with a random distribution of trees the method ensures that the more isolated trees are more likely to be sampled (see Chapter 2). If there is a patchy distribution the error is even greater. Some ecologists think that because of this non-randomness these plotless methods should not be used. The *T*-square sampling method can overcome this problem, and it is described in Chapter 2.

Plotless sampling is generally a much faster method for estimating plant density than frame quadrats in woods and forests, where the quadrats must be large to give a reasonable sample of the vegetation. For this reason, the technique could also be used in other sparse vegetation where individuals are distinct such as semi-desert or maquis-type vegetation. Each sample takes a longer time for the point-centred quarter method than for the nearest-individual method, but the latter gives a more variable estimate and you must carry out more samples to overcome this. The choice of method will therefore depend on the particular circumstances, including your preference. If plants are at very low density it may take a long time to find which is the closest plant. This is a particular, and at times overwhelming, problem if you are surveying the densities of individual species separately, especially since the rarer species are, by definition, at low density. If you sample species separately you must decide before surveying which species you wish to study. For this reason, you might not survey all the species in the vegetation, especially not the rarest. The more species you survey, the more work you will have to do, with a new distance measurement carried out at each sample point of each species. Where the community is species-rich, having many species with medium or low density (such as many undisturbed tropical forests), this may be a great problem. The method is probably best for surveys of overall tree density, of certain species of interest, or in species-poor communities.

Biases

Because sampling is non-random and different species may have different spatial distributions, a biased measure of relative densities will be obtained.

Seed-bank soil cores
Estimating the density of seeds in the seed bank

Methods

Soil cores of known area and depth are taken at sample points throughout the study area. Seed banks show great variation in seed density and species composition over small areas owing to the patchiness of species distribution, which can be exacerbated by the fact that the seed bank may be the accumulation of several years of seeding. You must therefore take many samples to achieve an adequate estimate. There is often great variation in the seed bank over time, reflecting temporal changes in seed production and germination, for instance owing to the seasons. You should either take samples on at least two occasions to assesss this variation or choose the sampling time carefully, for instance after the late summer peak in seed production in cool temperate zones.

If large enough the core can be dug out, using a frame quadrat (see p. 115) to mark the area. Smaller cores can be taken using a sheet-metal cylinder which has one sharpened rim and is pushed into the soil to a certain depth (Figure 3.3). It should then hold the core when it is

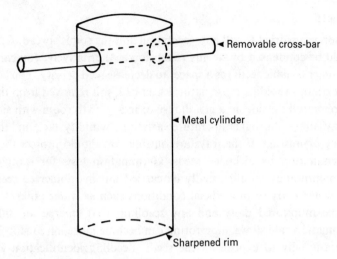

Removable cross-bar

Metal cylinder

Sharpened rim

Figure 3.3   A soil corer

removed from the soil. A removable cross-bar is useful to aid the pushing of the cylinder into the soil and to push the core out of the corer.

There is no standard core diameter or depth but generally diameter is in the range 2–20 cm and depth between 5 and 20 cm. The diameter should reflect the density of the seed bank and the size of seeds. Studies of arable fields and cultivated grasslands have used core diameters of 2–5 cm. The smaller the cores you use the more samples you will have to take in order to sample a reasonable area. The core depth should be decided after considering the aim of the study. Viable seeds are strongly concentrated in the top 2–3 cm of the soil, and these are the seeds most likely to recruit naturally into the community. The deeper-buried seeds will give a more complete picture of the seed bank and will also show the seeds available to recruit following soil disturbance.

The soil cores are transported back to the laboratory in bags to avoid seed loss or contamination. Here the core can be separated into layers, if you wish, to find the vertical distribution of seeds in the soil. A common division is into layers 0–2 cm depth, 2–5 cm and > 5 cm. The top layer will contain litter and you could scrape this off and analyse the seeds in this separately. In some studies the seeds of the litter layer have been considered not to be part of the seedbank and have been discarded. There is no ecological reason for this idea.

An alternative to soil cores in aquatic communities, such as lake beds, is to use a dredge grab which is set to sample to a specific depth, again between 5 and 20 cm. You should be able to roughly estimate the area sampled by the grab.

The seed bank in the cores can be estimated by germination tests or by counting seeds. Methods involving flotation of seeds on a saline density gradient are inaccurate and difficult and I shall not discuss them.

1. Germination tests

The core or core section is air-dried to kill any vegetation and the soil is spread over a seed tray. The sample could be condensed by sieving (see below). Alternatively, you can mix in sterile soil (i.e. containing no viable seeds) as a spacer to decrease seed density. You could use either a seed compost or, to provide a more natural seed bed, soil removed from the study site. The seed trays are placed outside, in a glasshouse or in a growth room with simulated natural conditions (day length, diurnal temperature variation, humidity, etc.) and the soil is kept moist by watering or misting. If the trays are outside you should protect them from herbivores and contamination by airborne seeds. Germination tests for samples from submerged aquatic communities should usually be carried out in submersed seed trays, either in the original water body or in artificial conditions such as water tanks.

The trays should be monitored daily and any seedlings that emerge identified and removed. As the germination rate slows, monitoring can become more infrequent. The soil should be stirred occasionally to expose all the seeds. Seedling identification keys are available for some species, but in the absence of expert help it is best to repot unidentifiable seedlings and grow them on until the adult plants can be identified. It could be useful to maintain a reference collection of seedlings. Monitoring can stop when nothing has emerged for several weeks.

An extra procedure which may increase the germination rate of some plants of temperate regions is to chill the seeds. The soil samples are placed in a fridge at about 5 °C for 3–4 weeks and then germination is tested in seed trays as described above. The warming after chilling simulates the return of spring after the winter conditions of the fridge and this is often a trigger to break seed dormancy.

2. Counting seeds

The seeds are sorted from the soil core by wet sieving. The sample is passed through a graded series of sieves, starting with the largest mesh size. Seeds will be trapped in different sieves according to the seed size. The smallest mesh size should be fine enough to catch the smallest seeds but coarse enough to allow the soil particles to pass through. A typical size is 150–200 μm. It is usually quicker to do this under running water and to break up soil aggregates with your fingers. Sieving can be carried out with sophisticated hydropneumatic elutriation equipment (e.g. Gross 1990) or simply with a collection of sieves and a tap. This method serves to concentrate the seeds, and the material remaining in each sieve should be sorted to find these seeds. You should therefore use a range of sieve sizes to speed up this sorting, so that the majority of seeds are not mixed in with too many stones and large particles.

The seeds are sorted from the remaining detritus, maybe using a binocular microscope for the smaller seeds, and are identified. You may be able to use identification keys or find expert help, or you may have to germinate the seeds and create a reference collection, as described above. It is usual to count how many of the seeds are still viable since the seed bank is defined only by the *live* seeds. To save time this can be done on a random subsample of the seeds to establish the proportion of the seeds of each species that are viable. Viability can be tested by germination tests in sterile soil or on dampened filter paper in a Petri dish or by tetrazolium and indigocarmine staining. Tetrazolium (triphenyltetrazolium chloride) stains living tissue red and remains colourless if the seed is dead. Indigocarmine remains blue if the seed is dead but goes colourless in live tissue. These tests are not always successful and so it is best to divide up the seed tissue and to carry out both tests using separate tissue samples. One positive result from the two tests is enough to indicate that the seed is viable. These stains are mildly poisonous so use them with care.

Tetrazolium

Fresh 1% 2,3,5-triphenyltetrazolium chloride is prepared in a phosphate buffer at pH 6–8. The seed is cut in half to expose its tissue and one half is incubated in the tetrazolium for 2 hours in complete darkness. Excess stain is washed away with distilled water. Living seed tissue will be stained a red colour.

Indigocarmine

0.05% indigocarmine is prepared in hot distilled water and the solution is filtered and allowed to cool. The seed tissue is submerged in the stain for 2 hours in the dark and then the stain is washed off with distilled water. If the tissue is colourless it is alive but a blue colour indicates dead tissue.

Advantages and disadvantages

Seed-bank sampling involves a lot of work no matter which method you use. Many samples and sampling on more than one occasion add to the effort. For aquatic samples, grabs are much quicker and easier than cores but there is much less accuracy in defining the area and depth of the sample.

Both methods for assessing seed numbers in the soil samples have drawbacks. It is virtually certain that not all the viable seeds will germinate in germination tests. While some species germinate readily others show seed dormancy which can only be broken by specific environmental factors. Chilling may break dormancy but other possibilities include after-ripening, scarification of the seed coat, high light intensity, widely fluctuating temperatures or even the severe heating of a fire. It is impossible to cover all the possibilities and most studies use just the standard conditions in the glasshouse or growth chamber. You must therefore remember that this method thus measures only the 'ready germinable fraction' of the seed bank. For these reasons it is also important to carry out germination tests in the same conditions for all samples.

By sieving the sample you will find most seeds, although the smallest seeds may be lost in the sieving and sorting processes. All stages in this technique are extremely labour intensive, and for this reason it is rarely used. The germination test for viability has the drawbacks described above. The chemical tests avoid these problems and provide the best method for fully sampling the seed bank, although they take a long time even if you use only a subsample.

Biases

The depth of soil core will probably affect the proportion of different species in your sample. Seeds are generally older further down the soil horizon and thus they represent both the longer-surviving species and the past composition of the plant community (which may be different to that at the present). The distribution of species in the soil can, however, be determined by separating the core into layers. The sample will also be biased towards those species that have produced seeds most recently. Sampling on different occasions will allow quantification of this bias.

The environmental conditions of the germination tests can strongly affect which species germinate. The sieving method may underestimate the very small-seeded species, such as the dust seeds of many of the Orchidaceae, since they may not be caught by the smallest mesh and may be too small to be seen in the sorting and sieving process.

Seed traps

Seed rain in terrestrial and non-submerged aquatic communities

Method

Seed traps are placed on the soil surface to estimate the density per unit time of seed arriving on that surface (the traps are sometimes placed vertically or at a slope but there is no benefit in doing this and it does not give a correct estimate of seed rain per unit area of the soil surface). A variety of trap types have been developed but sticky traps are by far the most popular. These prevent the loss of seeds from the traps. The basic premise of these is to fix a sticky surface to the ground which traps all seeds that come into contact with it. The sticky substance should be non-drying and not toxic to the seeds if they are to be germinated or tested for viability. There are a number of permanently sticky petroleum-based substances which can be used, for instance Tanglefoot, a bird repellant available in many countries and which can be smeared or aerosol-sprayed onto the trap surface. The sticky cards developed to sample flying insect pests can also be used. The surface used can be anything waterproof which can hold the sticky substance, for instance a pane of glass or polythene wrapped around wood or plastic. The insect cards could be used without modification. To be manageable the trap should have a diameter in the range 10–30 cm. Unless the trap is heavy you should fix it in place, for instance with long pins.

A popular alternative is to place a circle of paper (e.g. filter paper) smeared with the sticky substance in a Petri dish and use this as the trap. The dish can be either pinned to the soil surface or nailed onto a length of dowelling which you push into the soil to fix the trap in place. If you do the latter, the trap could be used to sample the seed rain onto the water surface in aquatic communities by fixing the dish above the water surface. You should make small holes in the dish to allow rain water to escape.

The trap should be removed or replaced after a few days or weeks. The period will depend on the density of the rain; if you leave it for too long, the seeds will be difficult to sort and the trap will become clogged up, reducing its efficiency. If you wish to carry out long-term sampling of the seed rain it would be wise to use a trap design which allows you to remove and replace the sticky surface easily. The Petri-dish trap allows you to substitute a new paper circle for the old one very simply.

The seeds are removed from the traps using forceps or a non-toxic solvent and are counted and identified. Viability can be tested using the staining or germination techniques described under the methods for estimating seed banks. If you germinate the seeds then clean off the sticky substance to allow water penetration. A simpler technique is also available for fresh seeds: non-viable seeds are usually partly or wholly empty and are therefore wrinkled or easily squashed. You should, however, test the accuracy of your judgement with the staining test. This method should not be used on older seeds because they may be intact yet dead.

As an alternative to sticky traps you can use seed trays containing a sterile loam or compost, the soil serving as the trap. The seedlings that germinate in the tray are identified and this gives a measure of the seed rain.

Advantages and disadvantages

All methods for assessing the seed rain are labour intensive. As with the seed bank the rain is very patchy so a good sample size is needed. The seed-tray method involves the least work but has the problems of germination tests described under 'Seed-bank soil cores'. However, if you are also studying the seed bank then you can use these germination tests to get a straightforward comparison of the rain and the bank.

Seed rain is highly variable over time (e.g. season), and if you want an estimate of the yearly seed rain, rather than that at a particular time, you must sample all year round (although in cool temperature zones you may be able to miss out the winter).

Ground wind may blow seeds and detritus over the soil surface into a sticky trap, causing an overestimate of the rain and clogging up of the trap. A shallow rim on the trap (such as that on a Petri dish) will avoid this, although too high a rim will block seeds falling at an acute angle onto the trap.

Biases

Very small seeds will almost certainly be missed if you sort and count the sample, and germination test have the biases described under the methods for estimating the seed bank.

Marking and mapping individuals

The performance of individual plants over a period of time in any vegetation

Method

Individual plants in the study area are marked and/or mapped so that they can be relocated and recognised at a later date. This is usually used in investigations on single species, such as demographic studies. You can study either some or all of the individuals in the study area. If you do the former you should either use frame quadrats (see p. 115) to delimit sample areas or choose individuals at random or according to a defined protocol (e.g. only flowering individuals). Mapping and marking can be carried out at any scale; from hundreds of km^2 for a rare rainforest tree species to 100 cm^2 for the tillers of a dominant pasture grass. In most cases it is best to both map and mark the individuals, either as a check that the mapping method has found the correct individual or as an aid to finding the marked plants.

The mapping method you use should depend on the scale of the study and on the density of individuals. You could simply note down the position of the plants on a map, locating them by coordinates (a global positioning system could be used; see vegetation mapping, p. 132) or landmarks. This is best for scattered individuals over a large area. Sample areas must be relocated and it is best to use 'permanent quadrats' for this. These are frame quadrats (see p. 115) which are fixed in place (e.g. using pins or tent pegs) or which can be replaced onto fixed markers, e.g. corner pegs. A useful method for smaller quadrats is to sink plastic or metal tubes (metal tubes can be relocated using a metal detector, but check beforehand that the metal detector can detect the tubes) into the ground and to attach legs onto the quadrat corners, which can then be placed into these tubes. Plants can be mapped within the quadrat by fixing a scale onto the frame or dividing the quadrat into a grid (see Frame quadrats, p. 115) and determining coordinates in relation to the frame using rulers or measuring tape. If the quadrat is small enough (i.e. <1 m^2) and the vegetation is short then you can trace the outlines of individuals using a pantograph (Figure 3.4), by using the grid as a guide to sketch by eye from the quadrat, or by using a 'mapping table'. This last is a quadrat modified so that it has legs that hold it above the ground and it holds a clear pane of plastic, perspex or glass. Tracing paper or a transparent plastic sheet is placed on this table and, looking straight down to avoid parallax, the outlines of individuals can be sketched directly onto the paper or plastic. Consecutive maps are superimposed to identify the individuals.

Individuals can be marked by a wide variety of methods; the best depends on the needs of the study, the plant size and structure and the materials available. If you want to distinguish every individual then they should be numbered. Numbers can be painted on the bark of slow-growing trees, written onto posts, canes, plant labels or toothpicks (the size of this 'post' depends on the plant size) placed next to the plant, or put onto horticultural plant tags, plastic-coated wire rings or even bird rings which you attach to the plant. You may wish to distinguish only groups of individuals, such as age cohorts. In this case you will not need to

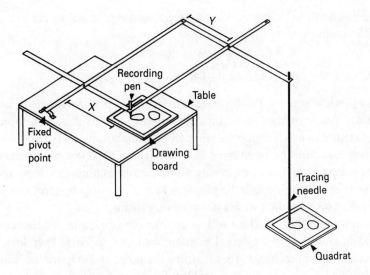

Figure 3.4 A pantograph. This can be either bought or made of wood, metal or plastic, although tubular aluminium with a plastic or plywood table are the best materials. You guide the tracing needle around the outlines of plants or point out the position of each plant. The recording pen inscribes these positions on a chart pinned out on the recording table. The resulting map can be scaled relative to the quadrat size, either larger or smaller depending on your needs. The scale depends on the ratio of the lengths X and Y. If $X > Y$ then the map is smaller than the quadrat.

number the plants but can distinguish the groups by the colours of the markers. The posts and tags described above can be coloured and you can also apply different coloured paints directly onto the plants. The paint should be non-toxic, fast-drying and hard-wearing, for instance artist's acrylic poster paint. With care this can be used even on very small plants such as grass tillers. Only small amounts of paint should be applied and only onto surfaces not important for light capture or gas exchange; the plant base is a good spot. The paint may need retouching at subsequent censuses.

On successive surveys each surviving individual is found again and you can make measures of performance on every plant. If individuals are numbered then you can simply note the performance measures against each number on your datasheets. If you have mapped plants without marking them with individual numbers, you must either write the measures on your map, or identify the individuals by their coordinates or by their position on the map (determined by overlaying consecutive maps) while you are in the field. Therefore, unless you are looking simply at survival between censuses, which can be determined by comparing consecutive maps in the laboratory, you should mark the individuals in the field. Mapping the outlines of individuals will allow you to calculate their basal areas. If you are censusing and marking or mapping all individuals in the study or sample area, you will locate new individuals at each census which you can then mark or map. If you are using a sample of individuals then if you want to locate new recruits (e.g. seedlings or young plants), you must resample at each census.

Advantages and disadvantages

These methods will allow you to follow the fate or measure the performance (e.g. leaf, flower or seed production, change in basal area, clonal growth etc.) of individuals in a population; a technique that is central to much of population ecology, population genetics and community ecology. Marking, locating and identifying individuals can be very time-consuming and detailed work in a dense population, especially where the plants are small. Despite this you must always take great care to disturb the plants as little as possible because you may alter their survival and growth by the process of measuring them.

Permanent quadrats or their markers will move over time owing to soil movement (e.g. frost heave) and intentional or accidental interference from animals. Therefore, as time passes the position of a plant relative to the quadrat will change and mapped individuals may be lost or misidentified. This is only a real problem for dense populations. Certain types of markers may be lost through vandalism or interference from other animals. For instance, wire rings are lost very easily from grazed grass tillers. You should consider this problem and make the markers as permanent as possible.

If an individual is only mapped or if the marker is not fixed to the plant then if the plant dies and a new plant grows up in the same place between censuses, you might mistake this new individual for the old one. New individuals may grow through wire rings to create the same problem. You must decide on the likelihood of this happening in relation to the vegetation type (low in forests, high in fertile grasslands) and the frequency of censuses.

Biases

The loss of markers or poor mapping may underestimate the survival of individuals. Misidentification of new individuals for old, dead ones may lead to overestimation of survival.

Vegetation mapping
Estimating cover of vegetation types over a large area

Method

'Vegetation type' can have a variety of definitions for the purposes of vegetation mapping and provides a way of categorising areas of similar species groupings and/or plant growth form. Starting out with only a vague notion of what classification you will use will result in a waste of time and poor quality information. You can devise your own categories, for instance based on the dominant species or species combinations (e.g. *Calluna vulgaris/Erica cinerea*, *C. vulgaris/Ulex minor*, *Agrostis curtisii/E. cinerea* etc.) or coarser criteria (e.g. high mangrove, mixed mangrove, salt marsh, savanna, pasture etc.). Usually this will supply sufficient information. Great detail can be obtained by classifying types by statistical ordination of species survey data (see pp. 276–319 in Kershaw & Looney 1983). Many standardised classification systems are available, the most famous being that of the Zurich–Montpellier school (see pp. 245–75 in Kent & Coker 1992 and pp. 156–64 in Kershaw & Looney 1983 for a full discussion of vegetation classification). The detail you can obtain will be limited by the mapping technique you use.

A number of techniques are available if you want to map and measure the cover (see p. 112) of different vegetation types over a large area, i.e. from a few to thousands of km². The vegetation types used (see Introduction) should depend on your needs and the scale of the survey.

1. Ground mapping

You can use a scaled-up version of visually estimating cover in quadrats with the quadrats being a convenient way of dividing up the survey area into manageable chunks. The size of these chunks should be determined by the total area to be surveyed and the detail you require. They can be measured out with reference to landmarks or as geographical entities (e.g. a hillside or the area between two streams) or you can use squares based on the grid of latitude and longitude. These can be located using a map and landmarks or by using a global positioning system (GPS) receiver, a hand-held device which uses satellite output to locate you precisely.

Either the cover of each vegetation type is estimated or their boundaries are sketched onto a map. The boundaries can also be positioned precisely using landmarks or a GPS. The areas on the map can be measured by hand or by using computer digitizers or image analysis systems.

2. Remote sensing by aerial photography or satellite images

This is highly technical (except for the most basic aerial photographs) and a full description is beyond the scope of this chapter. What is more, the technology is developing extremely rapidly, especially for satellite imaging. Curran (1985) is a good introductory text but you should consult experts. I shall merely give a background to allow you to decide whether you may find these techniques useful.

(a) You could simply take a normal photograph from a plane or helicopter directly over the vegetation. There are other more technical and more accurate procedures requiring specialised equipment. One popular alternative is to take aerial photographs as overlapping pairs, i.e. two parallel runs of photographs, one inclined from the left and the other inclined from the right. When these are viewed through a stereoscope a very clear three-dimensional image is seen. Aerial photographs at a range of scales are commercially available in many countries (see Curran 1985). The photographs may be infrared, multispectral (i.e. colour) or monochrome and are divided into patches of different tone, texture and pattern which correspond in some way to the vegetation. These cover types are related to vegetation types by field surveys of representative areas. The boundaries of each cover type can then be transferred to maps and their areas measured using tracing paper, by projecting the image onto a map, using computer digitizers (for instance, those associated with geographical information systems), using image analysis systems or using specially designed transfer instruments such as Sketchmasters.

(b) Satellite imaging involves computer-driven interpretation of satellite images such as those from the Landsat Thematic Mapper. The resolution (i.e. the spatial detail) of these images is determined by the pixel size used. The pixel is the area of land from which a single spectral image is taken and can now be less than $100\,m^2$. The spectral image measures the intensity of light at a range of different wavelengths. Pixels or groups of pixels are classified into types based on these readings and these types are interpreted using field surveys. The vegetation types may be very coarse, e.g. forest, grassland etc.; but much greater detail may be achieved, e.g. differentiating conifer, deciduous and mixed woodland or wet heath, dry heath and bogs. These satellite maps can be transferred onto computer systems, including geographical information systems, and the areas measured. Useful references that give the background to satellite imaging techniques are Turner & Gardner (1990) and Haines-Young *et al.* (1993).

Advantages and disadvantages

To contradict my statement about the complexity of the remote-sensing techniques a simple method is to take a normal photograph from the air, identify the cover types by ground surveys, sketch or trace the boundaries onto a map and measure the areas by hand. The ground-mapping methods and the more complex remote-sensing techniques give more accurate measurements and you will probably have errors of perspective on the photograph owing to camera tilt or changes in ground relief, which may lead to miscalculation of areas. However, this is a quick, cheap and crude method for large-scale mapping of vegetation. At greater expense you could buy accurate aerial photographs and analyse them by the crude methods given above.

Ground mapping is cheap in terms of equipment, requires no technical expertise and can produce very accurate maps. However, it is very time-consuming compared to the other techniques and for this reason cannot be carried out over very large areas. You can cover large areas by remote sensing (satellites can cover thousands of kilometres) and you can survey areas which are impenetrable on land. Once a system is developed, analysis of remotely sensed images can be very rapid. Neither of the remote-sensing techniques has the resolution of ground mapping, and the cover types will not be as precise as can be achieved by a surveyor on the ground. These cover types may also be misidentified. These techniques can be expensive and time-consuming.

Biases

Different vegetation types may not be distinguished by remote sensing owing to similarities in appearance or in spectral image.

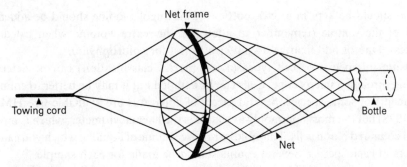

Figure 3.5 A plankton net. The towing cord can be replaced with a pole.

Phytoplankton

Density or volume of phytoplankton

Method

There are many methods for taking and analysing samples of freshwater and marine phytoplankton some of which involve specialised electronic equipment for *in situ* estimates. Analysis of samples can be by cell counts or indirect estimates of total cell biomass or volume (e.g. by chlorophyll *a* concentration). I shall describe only the simplest methods which use very little specialised equipment. I strongly recommend that you refer to the technical literature and consider other procedures (e.g. Bailey-Watts & Kirika 1981; HMSO 1983, 1984).

The water body can be sampled using a jar or container, although a plankton net can be used to cover a larger area and to concentrate the plankton sample. A plankton net (Figure 3.5) consists of a fine mesh net (e.g. 80 μm) ending with a bottle to hold the plankton. The net frame is usually square or circular, meaning that the sample area of the frame is easily calculated. The net should be small enough to be used easily and so the bottom of the water body is not disturbed (a problem especially in shallow streams). The net can be either fixed to a pole or to towing lines. For the latter the net can be towed either by you or by a boat. You can either hold the net against the current or sweep it through the water. The volume of water filtered depends on the net frame area, the speed of flow of the water, the speed you tow at and the time in the water. You should attempt to standardise this volume across all samples. Because of the fine mesh the net should be moved slowly through the water (>1 knot). The phytoplankton will be stratified (i.e. species show a non-random vertical distribution in the water body) so the net should either be held at the same selected depth in all samples or be moved evenly through all depths.

The algal cells will decay rapidly and should be preserved immediately after collection if you want to count them. Various fixatives are used but the commonest is Lugol's iodine solution. A solution of 20 g potassium iodide in 200 ml distilled water is made and then saturated by adding 20 ml iodine. This is then acidified with 20 ml glacial acetic acid. The solution should be kept in a dark bottle. 1–2 ml Lugol's iodine should be added to every 100 ml of the sample (remember to allow for this extra volume when calculating cell densities). Do not add fixative if you wish to test for chlorophyll *a*.

Densities of species or cell types (a less precise classification) can be determined by counting through a microscope. Cells can be counted but it may be better to count colonies or filaments of some species. Several pieces of apparatus are available (see HMSO 1990; Paert 1978) but the most commonly available is the haemacytometer, which is more usually used to count red blood cells. This holds a precise volume of liquid in which you can count the numbers of each species. Several counts should be made for each sample.

The concentration of chlorophyll *a* provides a relative measure of the volume of all phytoplankton in the sample. It does not distinguish species but can be useful for comparison of water bodies, for instance in studies of water quality. A number of methods are possible (see HMSO 1983; Riemann & Ernst 1982) but the simplest uses spectrophotometry. A defined volume of the sample is filtered (e.g. through a Whatman GF/C glass-fibre filter). You will need to carry out trial runs to determine the optimum volume to filter. The filter is then placed in 8–10 ml of 90% acetone and this is placed in a dark fridge for 20–24 hours to extract the chrorophyll *a*. The extract should be shaken two or three times during the incubation. Then the extract is removed from the fridge and allowed to warm up to room temperature in the dark. The extinction of the extract against that of 90% acetone should be measured in a spectrophotometer for the wavelengths 664 nm, 647 nm and 630 nm. Colloidal material in the extract will cause some turbidity and this can be corrected for by subtracting the extinction reading at 750 nm from these three readings. The relative concentration of chlorophyll *a* is calculated using the equation:

$$C_a = 11.85E_{664} - 1.54E_{647} - 0.08E_{630}$$

C_a is the concentration of chlorophyll *a* and E_i is the corrected extinction at wavelength *i*.

Advantages and disadvantages

These are the simplest and cheapest methods available. Netting phytoplankton will lose the smallest algae (nanoplankton) and other methods can be used to avoid this. Because algal cells vary widely in size (<1 to $>200\ \mu$m), counting cells does not give a complete picture of the community. You can also estimate cell sizes (see Bailey-Watts & Kirika 1981). Chlorophyll *a* degrades during extraction and some of the products are not detected by this technique. Some algae are very difficult to extract (e.g. blue-green) and may extract incompletely. These will cause an inaccurate estimate of chlorophyll *a*.

Biases

Poor control of the sampling depth of the sampling effort at each depth will bias the counts of different species. The smallest species will not be sampled by netting.

Benthic algae

Density or volume of benthic (bottom-dwelling) algae

Methods

Cover estimates can be made where benthic algae are large. Otherwise they should be sampled and the techniques described under Phytoplankton (p. 135) can be used to count cells or estimate chlorophyll *a*. Frame quadrats should be laid out and the algae in these sample areas removed. The removal technique depends on the substrate. More technical procedures than those given below are available, especially for deep water (see Flower 1985; HMSO 1990). For chlorophyll *a* analysis the extract should be refiltered and washed through with 90% acetone when it is necessary to remove the sediment and stones. The filtrate and washings are analysed.

In mud, silt or sand the surface sediment should be drawn up using a large syringe. The sample is analysed as it is.

Stones and small rocks should be removed and then washed in freshwater to remove the algae. More tenacious algae can be removed by scrubbing with a brush (e.g. a toothbrush) or scraping with a scalpel. Lumps of algae should be broken up by shaking or with a spatula.

Bedrock and other immovable objects must be scrubbed or scraped *in situ* and the dislodged algae should be collected with a large syringe. Lumps should be broken up.

Advantages and disadvantages

Microscopic enumeration of cells in sediment will be difficult as small algae will be hidden by the particles. It is difficult to remove cells from substrates with larger particles (2–20 mm) and other methods should be used (see Flower 1985; HMSO 1990). Cells are certain to be lost in the scraping and brushing of rocks.

Biases

Certain species may not be counted. Smaller cells in sediments may missed and the more tenacious species will be less likely to be dislodged from rocks.

References

Austin, M. P. & Heyligers, P. C. (1989). Vegetation survey design for conservation: gradsect sampling of forests in north-eastern New South Wales. *Biological Conservation* **50**, 13–32.

Bailey-Watts, A. E. & Kirika, A. (1981). An assessment of size variation in Loch Leven phytoplankton: methodology and some of its uses in the study of factors influencing size. *Journal of Plankton Research* **3**, 261–282.

Curran, P. J. (1985). *Principles of Remote Sensing*. Longman, London.

Flower, R.J. (1985). An improved epilithon sampler and its evaluation in two acid lakes. *British Phycological Journal* **20**, 109–115.

Goodall, D.W. (1952). Some considerations in the use of point quadrats for the analysis of vegetation. *Australian Journal of Scientific Research Series B* **5**, 1–41.

Grant, S.A., Suckling, D.E., Smith, H.K., Torvell, L., Forbes, T.D.A. & Hodgson, J. (1985). Comparative studies of diet selection by sheep and cattle: the hill grasslands. *Journal of Ecology* **73**, 987–1004.

Greig-Smith, P. (1983). *Quantitative Plant Ecology*. 3rd edn. Blackwell Scientific Publications, Oxford.

Gross, K.L. (1990). Methods for estimating seed banks. *Journal of Ecology* **78**, 1079–1093.

Haines-Young, R., Green, D.R. & Cousins, S.H. (1993). *Landscape Ecology and GIS*. Taylor & Francis, London.

HMSO (1983). *The Determination of Chlorophyll a in Aquatic Environments 1980*. Her Majesty's Stationary Office, London.

HMSO (1984). *Sampling of Non-planktonic Algae (Benthic Algae or Periphyton) 1982*. Her Majesty's Stationary Office, London.

HMSO (1990). *The Enumeration of Algae, Estimation of Cell Volume and Use in Bioassays 1990*. Her Majesty's Stationary Office, London.

Kent, M. & Coker, P. (1992). *Vegetation Description and Analysis*. Belhaven Press, London.

Kershaw, K.A. & Looney, J.H.H. (1983). *Quantitative and Dynamic Plant Ecology*. Edward Arnold, London.

Paert, H.W. (1978). Effectiveness of various counting methods in detecting viable phytoplankton. *New Zealand Journal of Marine and Freshwater Research* **12**, 67–72.

Riemann, B. & Ernst, D. (1982). Extraction of chlorophylls *a* and *b* from phytoplankton using standard extraction techniques. *Freshwater Biology* **12**, 217–223.

Turner, M.G. & Gardner, R.H. (1990). *Quantitative Methods in Landscape Ecology*. Springer-Verlag, New York.

Warren-Wilson, J. (1960). Inclined point quadrats. *New Phytologist* **59**, 1–8.

4 Invertebrates

Malcolm Ausden

School of Biological Sciences, University of East Anglia, Norwich NR4 7TJ,
United Kingdom

Introduction

Invertebrates, by virtue of their small size, are able to exploit very small and specific features within the environment. These features are known as microhabitats. Many invertebrate taxa, especially insects, also occupy different microhabitats during different stages of their life-cycle, and are consequently only present in these for limited periods during the year, e.g. a number of beetles and other insects spend their larval stages in runs of sap oozing from damaged trees. Because of this, it is frequently necessary to devise sampling strategies for invertebrates on a much finer scale than those used for many vertebrates. This will mean that when surveying invertebrates at a particular site (e.g. a woodland) it will usually be necessary to sample a wide range of different microhabitats within that woodland (e.g. dead wood of different tree species at different stages of decay and of different moisture content, the leaves of a variety of different tree and shrub species, wet and dry leaf litter, soil, bare ground, etc.). It may also be necessary to sample on a number of occasions throughout the year, in order to obtain a representative selection of species present. Conversely, when comparing invertebrates between sites, or at the same site over time, it is necessary to ensure that the fauna of similar microhabitats are being compared, and that they are sampled at similar times of year.

A vast array of trapping methods have been devised for sampling invertebrates. Many of these are large and expensive pieces of equipment and many consist of modifications of existing trap designs. The methods described in this chapter have been chosen to represent cheap, easy to use methods that are likely to be of more general practical use for invertebrate sampling (Table 4.1).

The activity of most invertebrates, especially insects, is often strongly influenced by weather conditions and the time of day. The level of activity may determine in which habitat or microhabitat a particular individual is at any one time (e.g. whether visiting nectar sources or resting in tall vegetation), how easy the individual is to locate, how easy it is to catch, and how likely it is to enter a trap. When comparing invertebrate faunas between sites, or at the same site over a period of time, it will usually be necessary to standardise the weather conditions and time of day under which the sampling takes place. This is of particular importance when using traps.

All trapping methods rely on invertebrates actively entering the trap. Catches of individuals within the trap will therefore reflect both the abundance and activity of the species, together with the species' susceptibility to being caught in the particular trap. Furthermore, catches in

Table 4.1. *The use of different methods for different groups*
* method usually applicable, + method often applicable, ? method sometimes applicable. The page number for each method is given.

Method	Page no.	Sponges, sea anenomes, hydroids, corals, starfish, sea urchins, sea cucumbers & other echinoderms, octopuses & other slow-moving or sessile taxa	Jellyfish, comb jellies & squids	Planarians	Earthworms & other terrestrial oligochaete worms	Aquatic oligochaete & polychaete worms	Leeches	Spiders & harvestmen
Direct searching	145	*	*	*	*	+	*	*
Water traps	150							
Flight interception traps	151							
Light traps	153							
Sugaring	154							
Aerial attractant traps	155							
Emergence traps	156							
Soil sampling	158				*			
Chemical extraction	159				?			
Separation of invertebrates from soil etc.	160				*			?
Pitfall traps	162							*
Beating	164							*
Sweep netting	165							*
Suction sampling	165							+
Pond nets & tow nets	167		*					
Cylinder samplers	169							
Robertson dustbin sampler	170							
Bait traps	170			*			*	
Digging & taking benthic cores	172			?		*		
Corers for use in fast-flowing water	174			?			+	
Kick sampling	174			?			+	

Method	Page no.	Pseudo-scorpions	Scorpions & other arachnids	Crabs, lobsters & crayfish	Water fleas, copepods & other zooplankton	Other aquatic crustaceans	Woodlice & other terrestrial crustaceans, millipedes, centipedes & other myriapods	Springtails, bristletails, thrips & booklice	Adult mayflies & stoneflies
Direct searching	145	+	*	*	+	+	*	+	+
Water traps	150								
Flight interception traps	151								
Light traps	153								
Sugaring	154								
Aerial attractant traps	155								
Emergence traps	156								+
Soil sampling	158								
Chemical extraction	159								
Separation of invertebrates from soil etc.	160	*					*	*	
Pitfall traps	162						+	?	
Beating	164							?	
Sweep netting	165							?	*
Suction sampling	165							+	
Pond nets & tow nets	167				*	*			
Cylinder samplers	169				*				
Robertson dustbin sampler	170								
Bait traps	170			*					
Digging & taking benthic cores	172					+			
Corers for use in fast-flowing water	174					?			
Kick sampling	174					?			

Table 4.1. (cont.)

Method	Page no.	Adult dragonflies & damselflies	Mayfly, caddisfly & stonefly larvae	Dragonfly & damselfly larvae	Grass-hoppers locusts & crickets	Mantises, stick insects & leaf insects	Cockroaches, earwigs & termites	Biting & sucking lice & fleas	Bugs	Adult beetles	Beetle larvae
Direct searching	145	*	+	+	*	+	*	*	+	*	*
Water traps	150										
Flight interception traps	151								?	+	
Light traps	153				?				?	?	
Sugaring	154										
Aerial attractant traps	155										
Emergence traps	156										
Soil sampling	158										+
Chemical extraction	159										
Separation of invertebrates from soil etc.	160						*				
Pitfall traps	162						+		+	+	*
Beating	164				?	*	+		*	*	+
Sweep netting	165				*	+			*	*	+
Suction sampling	165									+	?
Pond nets & tow nets	167		*	*					*	*	+
Cylinder samplers	169								*	*	+
Robertson dustbin sampler	170										
Bait traps	170		?	?					+	+	
Digging & taking benthic cores	172										
Corers for use in fast-flowing water	174		*	+							
Kick sampling	174		*	+							

Method	Page no.	Adult lacewings, antlion & other Neuroptera	Adult butterflies	Adult moths	Butterfly & moth larvae	Adult flies	Fly larvae	Adult sawflies & parasitic Hymenoptera	Ants	Bees & Wasps	Slugs & terrestrial snails	Aquatic molluscs
Direct searching	145	+	*	+	*	*	*	*	*	*	*	*
Water traps	150					*		*	*	*		
Flight interception traps	151											
Light traps	153	?		*		?		*		+		
Sugaring	154			+								
Aerial attractant traps	155		+			+						
Emergence traps	156					+	+					
Soil sampling	158						+					
Chemical extraction	159						?					
Separation of invertebrates from soil etc.	160					+	*				+	
Pitfall traps	162					+	*		+			
Beating	164	*		+	+						?	
Sweep netting	165	*			+	*	*	*			?	
Suction sampling	165					*		+				
Pond nets & tow nets	167						+					*
Cylinder samplers	169											
Robertson dustbin sampler	170											
Bait traps	170											
Digging & taking benthic cores	172						+					+
Corers for use in fast-flowing water	174						+					
Kick sampling	174						+					

traps that use attractants (light traps, sugaring, aerial attractant traps, baited pitfall traps, and bait traps) will be further biased towards species lured by the particular attractant. In these cases catches will also reflect the distance over which the attractant operates, and this may vary under different weather conditions and in different trap locations. Hence trapping methods provide indices of abundance and therefore cannot be used to estimate absolute abundance. It is important to be aware of this when interpreting results.

The difficulty of identifying many invertebrate species, together with the need to prevent invertebrates once caught in traps from devouring each other or dying and decaying, often requires them to be killed and preserved. Any trapping programme should take into account the likely effect that such removal of invertebrates may have on local populations. This is particularly important in the case of trapping large sexually mature invertebrates such as dragonflies, butterflies, and crickets, where the colony may only include a small number of adults. The effect of trapping on taxa that are not of relevance to the particular study should also be considered.

Temporarily storing live invertebrates

Live invertebrates should ideally be stored in glass or plastic specimen tubes or jam jars. If these are not available, clear plastic bags can be used.

A small amount of vegetation should be added to containers holding terrestrial invertebrates, so that they have something to attach themselves to during transport. Most insects, especially those with uncovered wings, should be kept in dry conditions, so that they do not become stuck to the sides of the container. Condensation within the container can be absorbed by placing a piece of crumpled tissue paper inside. Taxa taken from moist environments should be stored with a small amount of damp soil or vegetation.

Aquatic invertebrates should be kept in containers part-filled with water taken from where they were found. Freshwater species should never be stored in brackish or salt water or vice versa. Aquatic invertebrates, especially those from fast-flowing streams, are very sensitive to overheating, so should never be kept in containers in direct sunlight. Samples of mud or other benthos should be kept in sealed plastic bags in a refrigerator (below 4°C) and sorted, preferably within four days and at least within a week. Otherwise the invertebrates will die, decay, and become unrecognisable.

Containers should be kept in a cool, dark place to reduce stress on their occupants. Individuals of predatory species should obviously be stored separately.

Invertebrates should be released in the same place that they were caught, and in the case of groups such as bees and other nectar- and pollen-feeding insects, near flowers so that they can feed immediately if exhausted.

Killing and preserving invertebrates

All insects and hard-bodied invertebrates can be killed and preserved by dropping them into 70% alcohol solution. Although most other invertebrate groups can be adequately preserved in alcohol, many are better 'fixed' beforehand, particularly if they are to be used in reference collections. Fixation is the process of stabilising protein constituents in body tissue to help maintain them in a similar condition to that when the animal was still alive. A wide range of chemicals is used for 'fixing', and these are described in texts on individual groups.

Insects can also be killed by exposing them to ethyl acetate fumes. This is best done within a 'killing bottle'. This consists of a glass bottle (not plastic, as ethyl acetate corrodes this), containing a layer of set plaster of Paris onto which a few drops of ethyl acetate are dripped. Alternatively, ethyl acetate can simply be dripped onto a piece of crumpled tissue paper at the bottom of the bottle.

When using alcohol solution to store invertebrates, containers should be thoroughly sealed, since alcohol quickly evaporates. Cork stoppers are not suitable, because alcohol escapes through the pores in the cork. For storage of longer than a year it is advisable to add 5% glycerol to the alcohol solution to prevent specimens from becoming brittle or from completely drying out should all the alcohol evaporate.

Butterflies and moths should be preserved by pinning to prevent damage to the scales on their wings. Together with carding (gluing a specimen to a small piece of cardboard), this technique is often used to preserve insects required for display purposes. Details of these techniques are given in many entomology texts.

The best way to label specimens preserved in alcohol is in pencil on a piece of card placed with the specimen in the alcohol solution. Labels attached to the outsides of containers almost invariably fall off eventually.

Invertebrates should only be killed where there is no suitable alternative method of censusing.

Direct searching

All invertebrate groups

Method

The easiest way to find many invertebrates is simply by looking for them in suitable habitats or microhabitats. Many terrestrial groups of invertebrates require sheltered, moist microclimates, and can be found under stones, logs, and bark, around the bases of plants, in crevices in wall and rocks, and in leaf litter, nests, strandline litter, dead and decaying fungi, dung, and carrion. When one searches under stones or logs, it is important to return the stones or logs to their original position, to prevent animals remaining underneath from becoming desiccated. Care should obviously also be taken to cause minimal disturbance to the habitat.

Tussocks of grass and other vegetation support large numbers of invertebrates, particularly during winter when many hibernate within them. Invertebrates can be collected by slicing the tussock off at root level using a bread knife and then shaking and cutting it open over a white surface (e.g. a sheet or photographic developing tray). Smaller invertebrates are best located

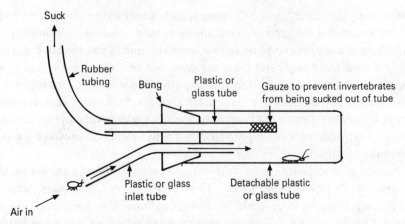

Figure 4.1   A pooter for picking up small invertebrates.

by taking tussocks indoors and sorting them under a strong lamp as described on p. 160.

Slugs and snails are more conspicuous during wet weather and especially at night when they can be found using a torch. Scorpions are best searched for using a torch with an ultraviolet-emitting bulb, since scorpions fluoresce when exposed to ultraviolet light.

More active invertebrates, particularly insects, may require more active search and capture by the surveyor, particularly the larger winged insects such as dragonflies, damselflies, butterflies, and moths. For flying insects it is often necessary to use a net. Wary fast flying insects such as dragonflies are easiest to catch in the early morning or in cloudy conditions when they are less active, although they may be harder to locate under these conditions. A pooter (Figure 4.1) or fine moistened paintbrush is useful for quickly and efficiently picking up small insects and arachnids. Pooters should never be used to suck invertebrates off dung or carrion because of the risk of inhaling bacteria or fungal spores.

Direct searching is also the simplest method for finding less active aquatic species. The most productive places to search are on and under stones (for mayfly, lacewing, alderfly, stonefly, and caddis fly larvae, and leeches), amongst aquatic vegetation (particularly in the axils of leaves of emergent plants, and in the sheathing leaf bases of tall monocotyledons), and amongst sticks and roots of marginal vegetation.

Invertebrates living amongst aquatic vegetation can be sampled by placing the vegetation in a shallow, white tray (e.g. a photographic developing tray) and removing invertebrates displaced from it. For greater effect, the tray can be painted with a large chess-board pattern of black and white, since some invertebrates are more easily seen on a black background than a white one. The greatest variety of invertebrates can be found by leaving the weed overnight in a water-filled, covered bucket. The subsequent oxygen depletion and slight fouling of the water encourages previously hidden invertebrates to come to the water's surface, where they are more easily visible.

On the seashore many invertebrates can be found by searching intertidal areas during low tide, especially during spring tides (i.e. at full and new moon) when the tidal range is greatest, enabling the lowest foreshore to be searched. The most profitable areas are amongst seaweed fronds, stems, and holdfasts and under stones in rock pools. Crabs are best searched for along the shoreline at night using a torch.

Many aquatic invertebrates are delicate and should be picked up using a small pipette or fine paintbrush, or failing that a piece of grass or small twig.

Mark–recapture is a suitable method for estimating populations of some invertebrate taxa with hard exoskeletons, and is discussed further in Chapter 2. The most widely used method is to mark the exoskeleton (avoiding joints or sensory organs) using an oil-based 'enamel' paint. This is most easily applied using the sharpened end of a matchstick. Other methods include marking the wings of butterflies and moths with felt tip pen after first rubbing a small patch of scales off, and gluing on individually numbered tags, for example, to the carapace of crabs.

In some cases it may be possible to standardise direct searching to some extent, in order to obtain relative population estimates. A number of examples are given below.

Counting numbers per unit effort

Timed searches are of more frequent use in aquatic habitats and have been used to make quick assessments of the invertebrate faunas of ponds. An example of such a method is to search each small (< 1 ha) pond for a total of three minutes, using hands and a net (p. 167), searching each habitat within the pond for a period of time in proportion to its area (Pond Action 1989). In terrestrial habitats the number of individuals counted in a set period of time has been used, for example, to obtain relative estimates of conspicuous taxa such as butterflies at different heights in the rainforest canopy (Hill *et al.* 1992).

Counting numbers of individuals per unit of vegetation

Invertebrates and plant galls can be searched for on individual leaves, stems (on monocotyledons), or entire plants. It is easiest to search leaves by checking both sides of consecutive leaves while moving from one end of a branch to the other. This can be made simpler by counting batches of, for example, ten leaves at a time, and then marking the position of the last leaf checked, using a clothes peg.

Alternatively, samples of foliage from trees or bushes can be carefully cut (most easily done using a branch cutter) and placed inside a large polythene bag, such as a plastic bin liner, taking care not to dislodge invertebrates on the foliage whilst this is being done. The sizes of the samples can be standardised by either (a) counting the number of leaves or buds per sample, after they have been checked for invertebrates or, (b) weighing the shoots or foliage. The first method is more suitable for plants with well-defined leaves, and the second for those with small or ill-defined leaves, such as conifers.

Once indoors, the foliage should be removed from the bag, the bag quickly resealed, and the foliage laid out on a large white sheet (or pieces of newspaper) on the floor. Invertebrates can then be dislodged from the vegetation by tapping it against the sheet, and then caught using a pooter. The foliage should then be carefully searched for invertebrates still remaining on it. Finally, the bag originally containing the sample should be carefully re-opened and any invertebrates inside it removed.

Densities of invertebrates and/or galls within particular parts of plants can be obtained by removing the required parts and dissecting them. This can be used to obtain estimates of invertebrate densities in, for example, flower-heads (e.g. Volkl *et al.* 1993) or stems of plants such as Reeds *Phragmites australis* (e.g. Ditlhogo *et al.* 1992).

Counting numbers of individuals per unit area

Conspicuous invertebrates can be surveyed by thoroughly searching for them within a defined area, for example, searching for male dragonflies holding territory over ponds (Moore 1964). Alternatively, smaller areas can be defined within the habitat by, for example, placing random quadrats throughout it. These can be searched immediately for relatively immobile taxa, such as mussels or buried cockles, or for the casts of polychaete worms or for holes made by the syphons of some bivalve species. Worm casts are usually formed just after the tide has receded; they then collapse at varying rates, depending mainly on the moisture content of the substrate. Numbers of casts should therefore be counted within a standard time of the tide having uncovered the substate, for example within 3–4 hours. In the case of more mobile taxa, such as grasshoppers and crickets, it may be best to leave the quadrat undisturbed for a while before searching it, to allow individuals disturbed during its positioning to return. Another method is to use a quadrat with high sides (a 'box quadrat') to help confine individuals while they are being counted (e.g. Cherrill & Brown 1990).

Counting exuviae (the cast-off skins of insects) is a useful method for giving an index of the productivity of insects that produce conspicuous exuviae in areas that can be easily and thoroughly searched, such as dragonflies and damselflies. Exuviae should be collected as frequently as possible, and these collections should be made throughout the whole emergence period of the species. The precise timing of this will vary from year to year, mainly in response to weather conditions. In the cases of dragonflies and damselflies, exuviae collecting may be made easier by placing sticks in the water in easily accessible places, for use by emerging nymphs.

Counting invertebrates along transects

The use of transects to monitor butterfly numbers was developed at the Institute of Terrestrial Ecology (Pollard 1977) and is now used in a national butterfly monitoring scheme in the UK (Pollard 1979). Thomas (1983) has developed this basic transect technique to obtain quick estimates of the relative sizes of butterfly colonies for use in surveys. Transects can also be used to monitor populations of other large, conspicuous invertebrates such as dragonflies and damselflies (e.g. Moore & Corbet 1990, Brooks 1993) and have been used to estimate numbers of larval aggregations of some species of butterflies and moths (e.g. Thomas & Simcox 1982). In aquatic habitats, transects have mainly been used to obtain estimates of numbers of sessile invertebrates on the foreshore, such as limpets (e.g. Lasiak 1991).

The use of transects is discussed more fully in Chapter 2. Where transects recording invertebrates are being compared, it is essential that the transects are carried out at similar times of day and under similar weather conditions.

Advantages and disadvantages

In most cases direct searching will be the most useful method for surveying invertebrates, since it allows suitable habitats and microhabitats to be quickly and easily investigated. Direct searching has the advantage, compared with trapping methods, of being selective. It also allows captured individuals to be released unharmed after being identified. However, it will often be less efficient in terms of numbers of individuals caught per time spent in the field than trapping methods.

Counting invertebrates in the field is fast and does not require any equipment but it does require prior identification knowledge of those invertebrates expected to be found. Removal of vegetation from the field is obviously destructive, and may require the transport of large amounts of vegetation. It can also be time-consuming. Counting conspicuous invertebrates or larval aggregations is quick and easy but only suitable in the few cases where conspicuous invertebrate taxa occur at high enough densities to obtain meaningful results, but not high enough that counting becomes impractical.

Although searching for exuviae may be relatively quick, because this has to be carried out at regular intervals throughout the entire emergence period, the whole process can prove very time-consuming.

Variations in searching efficiency between people mean that in order to be comparable, searches need to be carried out by the same individual.

Biases

Direct searching is likely to locate the more visually obvious, active, and large species, although the difficulty in catching large, active flying insects may result in these being under-recorded. Small and cryptically coloured invertebrates are likely to be under-recorded.

Disturbance during searching may cause some more active insects to fly off, resulting in underestimation of their numbers. Where foliage is removed, invertebrates may escape when the foliage is being sorted through. Mobile invertebrates may be missed or counted more than once.

Exuviae can be dislodged by rain or wind, leading to an underestimate of the productivity of the site.

Counts of numbers of casts or holes will to some extent reflect differences in activity as well as abundance of the particular taxa. Variable rates of collapsing of worm casts may bias counts. Molluscs buried in the substrate may move deeper, and hence not be counted, at different times of year and under different weather conditions (e.g. during cold weather).

Biases associated with the use of transects are discussed in Chapter 2.

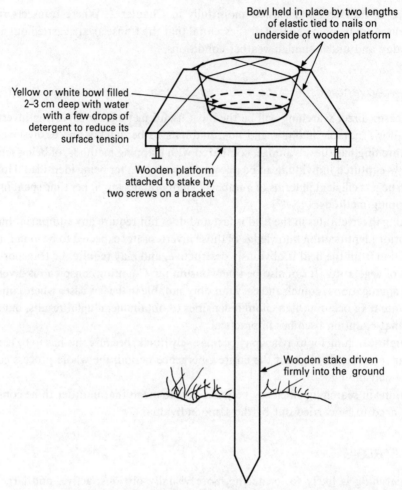

Bowl held in place by two lengths
of elastic tied to nails on
underside of wooden platform

Yellow or white bowl filled
2–3 cm deep with water
with a few drops of
detergent to reduce its
surface tension

Wooden platform
attached to stake by
two screws on a bracket

Wooden stake driven
firmly into the ground

Figure 4.2   A water trap for attracting and catching small flying insects.

Water traps

Flying insects, mainly flies and Hymenoptera

Method

Many flying insects are attracted to certain colours and can be attracted to and caught in coloured water-filled bowls (Figure 4.2). Yellow bowls are the best for catching both flies and Hymenoptera. White also attracts flies, but has a strong repellent effect on Hymenoptera. Alternatively, if 'neutral' coloured bowls, such as brown, grey, or blue are used, then these will have the least attractant (or repellent) effect on insects, and so reduce the selectivity of the sampling (Disney *et al.* 1982, Disney 1986, 1987).

The species composition of water trap catches varies with the height of the trap. Therefore, if being used to survey an area, a number of traps should be set at different heights to catch a wide range of species. Conversely, if being used to compare catches between sites, or at the same site over time, the height that the trap is set above the vegetation should be kept constant. Total trap catches are highest when the trap is just above the level of surrounding vegetation (Usher 1990).

When one is sampling in woods it may be necessary to fix a wide-mesh gauze over traps to prevent leaves from falling in and affecting the attraction of the trap.

Traps should be emptied at least once a week. Invertebrates can be removed from the traps by pouring the contents through a piece of muslin into a bowl. The muslin containing the specimens can then be removed and placed in a collecting tube containing 70% alcohol solution, and the water trap liquid returned to the trap.

Advantages and disadvantages

Water traps can be used to sample invertebrates in virtually all habitats. Traps have to be emptied at frequent intervals, otherwise their contents will decay unless a preservative is used. Preservatives will affect the attractiveness of the trap to flying insects. Traps also need to be constantly checked to prevent them from overflowing or drying up.

Water traps are virtually impossible to protect from grazing stock, particularly cows, which use them as drinking troughs. Fencing off the traps from stock is likely to affect the vegetation in the immediate vicinity of the trap, thereby affecting its catch. Traps are also fairly conspicuous and are therefore likely to be disturbed by passers-by. Insects caught in water traps are sometimes eaten by small birds, which once having learnt of this rich source of food, may regularly remove large numbers of insects from the traps.

Biases

As with all trapping techniques, numbers of insects caught in the traps will depend on their activity and their attraction to the colour of the traps, as well as their abundance. Variable rippling of water in the traps caused by wind will also affect the trap's attractiveness.

Flight interception traps

Flying insects

Method

Flight interception traps work by blocking flying insects with a screen of fine black netting. Blocked insects then drop down into collecting trays laid beneath the netting (as in the type shown in Figure 4.3), or are guided upwards into a collecting bottle (in the case of Malaise

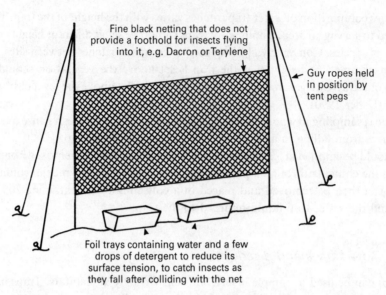

Fine black netting that does not provide a foothold for insects flying into it, e.g. Dacron or Terylene

Guy ropes held in position by tent pegs

Foil trays containing water and a few drops of detergent to reduce its surface tension, to catch insects as they fall after colliding with the net

Figure 4.3   A flight interception trap for intercepting and trapping flying insects.

traps). Malaise traps are complicated and expensive structures and are not specifically dealt with in this chapter.

Traps are best sited in areas frequented by large numbers of flying insects such as along woodland edges or woodland rides, or near hedges. Flight interception traps can be checked as infrequently as once a week.

Advantages and disadvantages

Traps are large and conspicuous and therefore prone to disturbance from passers-by. Flight interception traps are rarely used to compare numbers of insects between sites or at the same site over time, because their size tends to make replication impractical.

Biases

Flight interception traps, such as the one illustrated, catch few small active insects, and proportionally more heavy, cumbersome flying insects, such as larger beetles. Malaise traps are better at catching smaller, more agile flying insects, particularly flies and small Hymenoptera. As with all trapping methods, the catches will reflect both the abundance and activity of particular species.

Light traps

Mainly moths, but also other night-flying insects

Method

Many night-flying insects, particularly moths, are attracted towards light, particularly that at the ultraviolet end of the spectrum. They can then either be actively caught, or encouraged to enter a trap.

The simplest light trap consists of a light on a cable (an inspection lamp is ideal) hanging outside a building. Any bright white or bluish light is suitable, although a high-pressure mercury vapour bulb is best. Most mercury vapour bulbs work off direct current, so an adaptor is necessary to run the lamp off the mains. As well as being very bright, mercury vapour bulbs also emit ultraviolet light. The effectiveness of the trap can be enhanced if the lamp is positioned beside a white wall or has a white sheet hung next to it. Alternatively the light can be hung from a tripod above a white sheet, in which case the tripod must be earthed. Seek the advice of a qualified electrician if in any doubt. Light traps should not be operated during rain, because water falling on the hot bulb will crack it. If no electricity supply is available then a paraffin vapour or gas lamp can be used instead.

Catches of moths can be increased by using a moth trap, such as that shown in Figure 4.4. These can be left running for long periods without the observer needing to be present.

Traps should normally be situated so that their light source is not heavily shaded by surrounding foliage, since this will reduce the area over which the light is visible to flying insects. Light traps attract most moths on warm, overcast nights with little wind, and especially when it is thundery.

If live moths are kept for examination, then afterwards they should be released at dusk. If this is not possible, then they should be released under cover and not on lawns or bare surfaces where they will be easy prey for birds.

Advantages and disadvantages

Light traps are capable of catching very large numbers of moths under favourable conditions. For example, during a trapping programme in Kenya, one trap caught 49 000 moths in one night! (Brown *et al.* 1969). Catch rates using light traps are very variable according to the weather conditions and so cannot be used to monitor moth numbers, unless used virtually every night over a period of years.

Biases

Light trap catches are biased towards species attracted to light and species most likely to enter and remain in the trap. As with all trapping techniques, catches reflect the activity of the individual species.

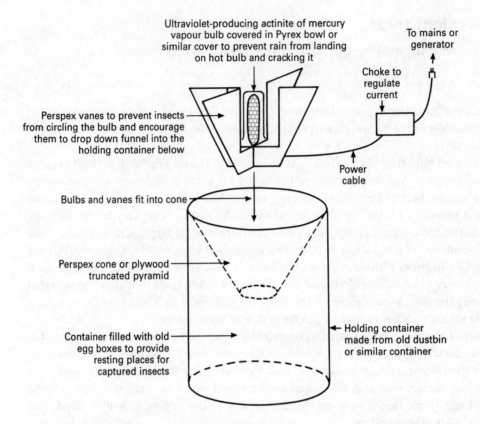

To mains or generator

Ultraviolet-producing actinite of mercury vapour bulb covered in Pyrex bowl or similar cover to prevent rain from landing on hot bulb and cracking it

Choke to regulate current

Perspex vanes to prevent insects from circling the bulb and encourage them to drop down funnel into the holding container below

Power cable

Bulbs and vanes fit into cone

Perspex cone or plywood truncated pyramid

Container filled with old egg boxes to provide resting places for captured insects

Holding container made from old dustbin or similar container

Figure 4.4   A light trap for attracting and catching night-flying insects.

Sugaring

Moths

Method

This method consists of attracting moths to a sugary solution laced with alcohol, which renders them less agile.

Sugaring attracts the largest number of moths during warm, overcast weather. The basic 'sugar' consists of black treacle (or Barbados or molasses sugar) boiled up with over-ripe bananas or other rotting fruit. Just before use, some alcohol (e.g. beer) should be added, and (optionally) a couple of drops of amyl acetate to give a strong scent to the mixture.

The 'sugar' should be applied at dusk on foliage, tree trunks, concrete, or wooden posts using a stiff 2–3-cm-wide paint brush or bare hands. Sheltered spots are best if it is windy. The 'sugar' should be revisted every hour. A weak torch can be used to search for moths in the immediate vicinity as well as on the 'sugar' itself.

Advantages and disadvantages

Sugaring requires the recorder to be constantly present. Since catch rates vary greatly with position and weather conditions, sugaring cannot be used to compare numbers of moths between sites or at the same site over time. Sugaring often attracts moth species that do not usually come to light.

Biases

Catches are restricted to the relatively few species attracted to the 'sugar'. Numbers caught will be a reflection of both their activity and the distance over which they are able to detect the 'sugar', and this will also vary with weather conditions.

Aerial attractant traps
Flies and butterflies

Method

Flies can be attracted into containers holding suitable baits, and then trapped within these or guided upwards into a collecting bottle (Figure 4.5). A wide range of baits can be used: rotting fruit for fruit flies, dung for dung flies, carrion for carrion-feeding species, carbon dioxide (in the form of 'dry ice') wrapped in polythene for slow release for blood-sucking species, fungi, fish, rotting eggs, etc. To obtain the widest variety of species when using baits that decay, such as meat, it may be worthwhile leaving bait in different traps for varying periods of time since different fly species are attracted to meat in different stages of decay. Butterflies in the tropics can be attracted to and caught in traps baited with rotting fruit (especially over-ripe bananas) or dung.

Advantages and disadvantages

Wasps and other larger Hymenoptera may enter traps, particularly those containing fruit, and kill and damage insects already caught inside. Ants also often enter the traps, especially in the tropics, and can kill and remove insects already caught.

Biases

Catches are biased towards flies that are attracted to the particular baits provided and are liable to enter and become caught in the trap, and will also be influenced by the species activity. Flies already caught in traps may attract other flies (Cragg and Ramage 1945).

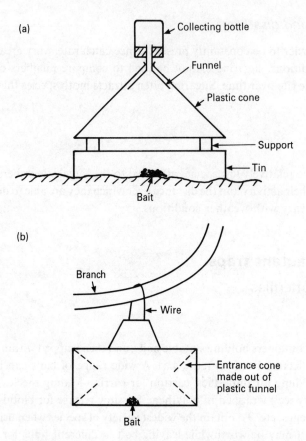

Figure 4.5   Aerial attractant traps. (a) and (b) are ground-based and hanging traps respectively, both used for catching flies (from Stubbs & Chandler 1978). (c) is a hanging trap used primarily for catching butterflies in the tropics.

Emergence traps

Mainly flies but also mayflies and caddis flies

Method

Emergence traps (Figure 4.6) can be used to trap adult insects emerging from their pupae in water, benthos, soil, dead wood, etc. The basic design of these traps is similar, and consists of a container that fits snugly over the substrate or water surface from where the adult insects emerge. This container usually has an opening to the light at its highest point, which the entrapped insects move towards. Insects caught in traps without collecting bottles can be removed by spraying with 70% alcohol solution using a fine household plant sprayer, or by holding the trap up to the light and removing the insects using a fine paintbrush dipped in 70% alcohol solution.

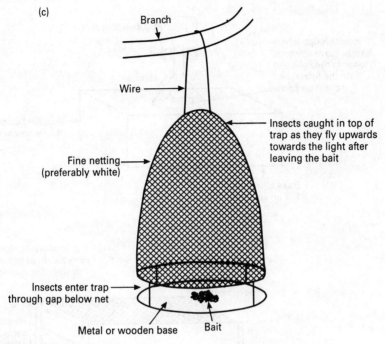

Figure 4.5 (*continued*).

Emergence traps will provide absolute or relative estimates of numbers of emerging insects, depending on their efficiency. To standardise losses, traps should be emptied at regular intervals.

Advantages and disadvantages

Aquatic emergence traps can be damaged by wave action, and are usually difficult to use at sites with widely fluctuating water levels. Terrestrial traps are often conspicuous and therefore likely to be disturbed by passers-by.

Biases

Larvae about to emerge may be attracted to the trap (owing to the favourable, sheltered conditions for emergence that it might provide), resulting in an overestimate of numbers of emerging insects, or may be repulsed (many larvae and pupae are attracted towards light and will therefore move away from any shadow cast by a trap), resulting in an underestimate. Numbers of insects caught will also depend on the frequency of collection. Even under very favourable conditions the proportion of insects lost from the traps will increase the longer the traps are left. Loss occurs mainly through decay and sinking, and the activity of predators, and will be increased by wave action. Reviews of the efficiency of emergence traps are given by Mundie (1971) and Morgan (1971).

(a)

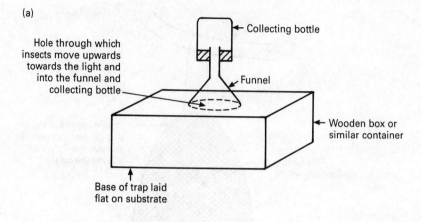

Hole through which
insects move upwards
towards the light and
into the funnel and
collecting bottle

Collecting bottle

Funnel

Wooden box or
similar container

Base of trap laid
flat on substrate

(b)

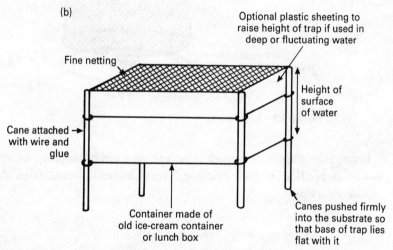

Optional plastic sheeting to
raise height of trap if used in
deep or fluctuating water

Fine netting

Height of
surface
of water

Cane attached
with wire and
glue

Container made of
old ice-cream container
or lunch box

Canes pushed firmly
into the substrate so
that base of trap lies
flat with it

Figure 4.6 Emergence traps for catching adult insects emerging from the ground (a) or from shallow water and benthos beneath it (b).

Soil sampling

Large soil invertebrates, especially earthworms, fly larvae, and beetle larvae

Method

This method involves taking soil samples of a known volume. These can be dug using a spade (a quick way of measuring the surface area of samples is to dig around the outside of a wire quadrat), or by using a corer where the substrate is soft.

The majority of earthworms and larger invertebrates such as fly and beetle larvae can be removed while breaking the soil up by hand. To recover less conspicuous individuals the sample can then be 'wet sieved' (see p. 160) using a mesh size of 2 mm. The extra proportion removed by additional wet sieving will depend on the nature of the soil and the sizes and visibility of the invertebrates being removed. In many cases it may be most efficient to hand sort the entire sample but to wet sieve only the root layer, which can be more difficult to sort thoroughly by hand and which usually contains a large proportion of the individuals within the sample, particularly small inconspicuous surface-living earthworm species.

To remove even smaller and less conspicuous invertebrates it is suggested that samples are hand sorted, wet sieved, and then sorted by flotation (p. 160).

Advantages and disadvantages

In general, sorting dug samples gives more reliable estimates of earthworm numbers and biomass than its main alternative, chemical extraction (p. 159). However, digging and carrying large numbers of soil samples is time-consuming and tiring. For a more detailed comparison of techniques of monitoring earthworm numbers see Nordstrom & Rundgren (1972).

Biases

The proportion of worms extracted using this method varies with soil type. Raw (1960) found that hand sorting using flotation recovered 52% of the total number of worms (84% of the total weight) in good-quality soils. Small and dark coloured worms are under-recorded compared with when using chemical extraction (e.g. Raw 1960, Nordstrom & Rundgren 1972).

Chemical extraction
Earthworms and tipulid fly larvae

Method

Formaldehyde solution is an irritant to earthworms and tipulid fly larvae and can be used to expel them from the soil. A solution of 0.2% formaldehyde is suggested. Extreme care should be taken when handling formaldehyde and a safety mask and gloves worn. The solution should be evenly applied, using a garden watering can with a sprinkler rose, to an area of, for example, $0.5 \, m^2$ of soil, so that the soil is thoroughly infiltrated without run-off. All invertebrates that emerge within a standard period of time (e.g. 10 minutes) should be collected. Care must be taken not to use too strong a solution, as this will kill invertebrates in the soil. Expelled invertebrates should be picked up (preferably using forceps, since the formaldehyde makes them secrete large amounts of mucus) and washed briefly in clean water, so that they can then recover.

Advantages and disadvantages

Chemical extraction is quick but suffers from the disadvantage that it is not possible to quantify the area, and in particular the depth, of soil affected by the solution. Hence it is not possible to calculate invertebrate densities per known volume of soil. The area of the soil that the chemical affects varies with soil moisture, and so this method cannot be used to compare invertebrate densities in soil of very different moisture content.

Biases

This method is inferior to hand sorting for extracting earthworm species experiencing periods of diapause (usually during dry periods of the year), since in this state they do not respond to the formaldehyde solution (Nordstrom & Rundgren 1972).

Separation of invertebrates from soil, litter, and other debris

Soil- and litter-inhabiting invertebrates

Method

Sieving can be used to sort invertebrates from any substrate that can easily be broken into smaller pieces. Dry material can be sieved onto a white cloth, piece of newspaper, or polythene, starting with a 3–4-mm-mesh sieve and working down to a 0.5-mm-mesh sieve. Some invertebrates such as beetles and pseudoscorpions become motionless when disturbed, so it is best to wait a short time before discarding sorted material. It is also useful to sieve under a strong light. Not only does this make it easier to see, but as some groups are negatively phototactic and are also stimulated by heat, they become more active under a strong light and hence are easier to find.

Very fine, and in particular wet, substrate, such as mud or sand from cores (see p. 172) is often more easily sorted by running a jet of water through it during sieving, using a short length of pipe attached to a tap ('wet sieving').

Invertebrates can be separated from soil, litter, etc. by stirring it up in a bucket of water and collecting the specimens as they float on the water's surface. If there is a large amount of litter in the debris in the sample, then it is necessary to stir continually and break this up to release invertebrates so that they are free to float to the surface. Small specimens are most easily picked up using a small pipette or paintbrush. This method is good for recovering beetles, but to make other invertebrates float it is usually necessary to increase the specific gravity of the water by adding sugar or salt. If live specimens are required then they should be picked off from the water's surface and washed in clean water as soon as possible.

If the solution is being used to compare numbers of invertebrates between sites, or at the same site over time, it is important that the strength of the flotation solution remains constant, so that the proportion of invertebrates made to float by the solution remains the

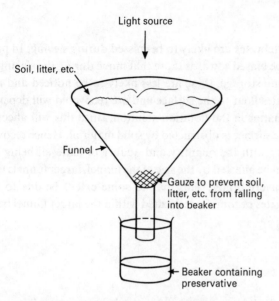

Figure 4.7 A Tullgren funnel for separating small invertebrates from soil, litter, etc.

same. This is most easily done by measuring the specific gravity of the solution using a beer- or wine-maker's hydrometer. Sieving the material to be sorted beforehand removes the larger particles, and makes the method more efficient.

Desiccation funnels, such as the Tullgren funnel (Figure 4.7) can be used to extract invertebrates from a variety of loose, large-particled substrates, such as soil, litter, flood debris, or old birds' nests. They work by creating warm, dry, and light conditions at the top of the funnel, which encourages cool-, shade-, and moisture-loving invertebrates to move down the funnel away from the light source, until they eventually fall out of the bottom of the funnel into a collecting bottle. If live specimens are required then a lightly moistened piece of filter paper should be placed in the collecting container. Funnels are usually left in operation for a week or so, and if live specimens are being collected, they should be checked daily.

A wide variety of different designs of this basic apparatus have been devised, including the 'Berlese funnel', which instead of having a light source above it, consists of a metal funnel encased on its outside by a jacket through which hot water is passed, and the 'Baermann funnel', used mainly for collecting nematodes and groups such as rotifers. These and other modifications to desiccation funnels are reviewed by Murphy (1962).

Advantages and disadvantages

Wet sieving and recovering invertebrates by flotation can be messy and time-consuming and either needs to be done outside, or requires a large sink that is not likely to become blocked by sediment in the run-off water. The use of desiccation funnels is not labour-intensive, since sorting can be left unattended.

Biases

Small and inconspicuous invertebrates are likely to be missed during sieving. In particular, species recovered are likely to be biased towards those that move during the sieving process. If invertebrates have died during storage, these are less likely to be noticed and recorded.

The rate at which invertebrates float to the surface and are recovered will depend on the quantity and nature of solid matter in the flotation solution, since this will affect to what extent their upwards path to the surface is obstructed by solid material. Hence recovery rates per time spent sorting will vary with the quantity and quality of material being sorted.

The catch from the funnel will be affected by the size of the funnel, larger funnels tending to extract relatively more large invertebrates. This may to some extent be due to a greater proportion of smaller invertebrates becoming desiccated within the larger funnel before they reach the collecting tube.

Pitfall traps

Active, surface-living invertebrates in low vegetation or bare ground

Method

Pitfall traps consist of straight sided containers sunk level with the surface of the ground into which invertebrates inadvertently fall (Figure 4.8). Any size or type of container with smooth sides is suitable. Traps with a larger circumference will catch more invertebrates, but this should be balanced against the need to set enough traps in different microhabitats to ensure that the catches are representative of the site as a whole.

A number of preservative solutions can be put in the trap to prevent invertebrates from eating each other and arrest decay. Several solutions make effective preservatives. Ethylene glycol (anti-freeze) has the advantage that it does not evaporate and so traps containing it can be checked as infrequently as once a month. It is potentially harmful to skin and eyes so protective gloves and eye wear are advisable. Availability can be a problem during summer months and in tropical countries. Staining may also be a problem if soft-bodied specimens are required for display purposes. Alcohol is benign, but has the disadvantage that it evaporates quickly, and so traps containing it need to be checked at least weekly, and even more frequently in hot and windy weather. Other alternatives are formaldehyde solution and commercial insecticides. Extreme caution should be exercised with formaldehyde, which is a very dangerous chemical. Commercial insecticides can also be harmful, so the manufacturer's instuctions should always be followed carefully.

The composition of catches will vary with the design and size of the trap, with the choice of preservative, and depending on whether or not the trap is covered (e.g. Greenslade & Greenslade 1971, Luff 1975). Hence these need to be standardised if catches are being compared between sites or at the same site over time.

Pitfall traps act as water traps (p. 150) for flying insects such as flies and Hymenoptera.

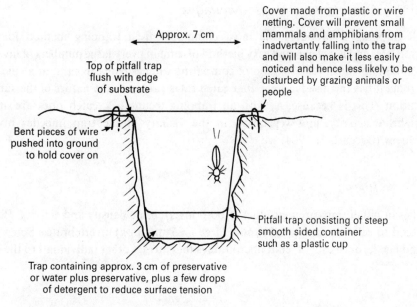

Approx. 7 cm

Top of pitfall trap
flush with edge
of substrate

Cover made from plastic or wire
netting. Cover will prevent small
mammals and amphibians from
inadvertently falling into the trap
and will also make it less easily
noticed and hence less likely to be
disturbed by grazing animals or
people

Bent pieces of wire
pushed into ground
to hold cover on

Pitfall trap consisting of steep
smooth sided container
such as a plastic cup

Trap containing approx. 3 cm of preservative
or water plus preservative, plus a few drops
of detergent to reduce surface tension

Figure 4.8   A pitfall trap for catching invertebrates moving on the surface of the ground or amongst low vegetation.

However, they should not be used to compare numbers of these groups since the catch will be strongly influenced by the colour of the pitfall trap and how visible it is. These factors will vary both between sites and over time at the same trap. Visibility will be heavily influenced by the nature of the surrounding vegetation, and the colour of the pitfall trap will actually change over time as more invertebrates are caught in it.

Pitfall traps may also be baited with raw meat, fish, cheese, fermenting fruit, etc. to attract beetles. Preservatives should not be used in baited traps as these may mask the smell of the bait. Such traps therefore have to be checked daily.

If traps are to be repeatedly emptied, then a quick way to do this is to empty the trap contents through muslin (see p. 151). The trap liquid can then be re-used, although it may be necessary to top it up with preservative, especially, in the case of alcohol, if it has evaporated or become diluted with rain water. After traps have been emptied it is worth wiping their inside surfaces with a cloth, to keep them clean and smooth (particularly if slugs and snails have entered and left behind mucus). This will maintain the catching efficiency of the trap.

It is easiest to place pitfall traps in a line or cross, to aid relocation. In order for catches in individual pitfall traps to be independent of each other, as a rule of thumb, traps should be set at least 2 m apart. In most cases it is sensible to mark the position of traps with a small post, since they can be surprisingly difficult to relocate, especially if left for long periods during the growing season, when they may become quickly obscured by vegetation. Marking traps with posts will, however, make them more conspicuous to passers-by, and to grazing stock, which may then damage them.

Advantages and disadvantages

Pitfall trapping is probably the most commonly used trapping method for studying invertebrates and is a cheap and easy method of catching very large numbers of invertebrates with minimum effort. As a method of comparing catches between sites, or at the same site over time, it has the disadvantage that catch rates vary with the nature of the surrounding vegetation. This is because, as with all trapping techniques, catch rates are affected by invertebrate activity, and vegetation in the vicinity of the trap impedes invertebrate movement (Greenslade 1964).

Biases

Catches in pitfall traps are a product of both invertebrate density and activity. Pitfall traps also tend to catch proportionally more large (> 3 mm long) invertebrates. Some species of ground beetle, once caught, emit pheromones that attract other individuals to the trap (Luff 1986).

Beating

Invertebrates on foliage

Method

Beating is a simple technique which involves sharply tapping branches with a stick and catching dislodged invertebrates in a beating tray held beneaath. A beating tray is a cloth-covered tray which is slightly sloping towards its centre. They can be bought or made, or alternatively an old umbrella or a sweep net can be used. Invertebrates can be collected from the tray using a pooter (p. 146).

Variations on the basic beating technique include tapping branches and netting the insects as they fly away, and shaking tussocks of vegetation in marshes over a white tray to collect marshland snails.

Advantages and disadvantages

Beating is a very quick and easy way to collect large numbers of invertebrates. It can be used to produce relative estimates of invertebrate numbers, for example by tapping individual branches a standard number of times.

Biases

Beating is biased towards species that are easily dislodged but which do not fly when disturbed.

Sweep netting

Invertebrates in low vegetation

Method

The method involves passing a sweep net through the vegetation using alternate backhand and forehand strokes. Nets need to have a reinforced rim. Round or kite nets are unsuitable. After completing a series of sweeps, invertebrates caught in the net can be encouraged to move to the closed top of the net by holding this end up towards the light. If specific invertebrates or invertebrate groups are being sought, then these can then removed using a pooter (p. 146). If all the contents are required then the contents of the net can be emptied into a killing bottle (p. 145).

Advantages and disadvantages

Sweep netting is a quick, low-cost, and efficient way of collecting large numbers of invertebrates, making it well suited for surveying purposes. However, sweep netting cannot be carried out if the vegetation is damp and does not work well in vegetation less than 15 cm high, or which has been flattened by wind, rain, or trampling. It is of more limited value for purposes of comparison monitoring, because of variations in the efficiency of sweep netting in differently structured vegetation. The catch obtained will also be influenced by the speed, depth, and angle at which the net is pulled through the vegetation. For example, many flies will avoid a slowly approaching net but be caught by a faster one. Hence in order for comparisons to be made, the mode of sweep netting should be standardised and the samples taken by the same person. An easy way to standardise the method is for each sample to consist of a series of net sweeps of approximately 1 m in length taken every other pace while walking at a steady speed through the vegetation.

Biases

Sweep netting tends to underestimate numbers of invertebrates that cling tightly to the vegetation (e.g. Lepidoptera larvae) and also species that take evasive action (either dropping to the ground or flying off), such as some grasshoppers and flies.

Suction sampling

Invertebrates in low vegetation

Method

Suction sampling involves the sucking up of invertebrates from a known area of vegetation into a net. The most commonly used purpose-built suction sampler is the D-vac, which is a

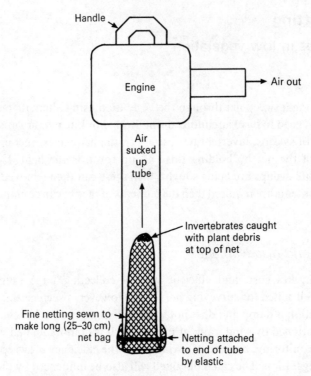

Handle

Engine

Air out

Air
sucked
up
tube

Invertebrates caught
with plant debris
at top of net

Fine netting sewn to
make long (25–30 cm)
net bag

Netting attached
to end of tube
by elastic

Figure 4.9   A suction sampler modified from a machine designed for removal of garden litter. This can be used to suck up invertebrates from low vegetation and bare ground.

large piece of apparatus, carried on a person's back. A smaller, lightweight sampler can be made by converting a suction machine designed for removal of garden litter (Figure 4.9). These have comparable, and in many cases greater, suction power (and hence collecting efficiency) than D-vac samplers (Wright & Stewart 1992). They are available at some garden centres. Suction samplers run off two-stroke engines.

Two methods of monitoring invertebrates using suction samplers can be used. The collecting nozzle of the sampler can be pushed vertically downward into the vegetation and held there for a standard length of time (e.g. ten seconds) to suck up invertebrates from an area of vegetation the size of the sampler's nozzle (many samplers may also suck up invertebrates from outside this area, particularly as they are being lowered or raised). This can then be repeated a number of times. It is necessary to keep the sampler running between individual sucks to prevent collected specimens from escaping. Alternatively, a known area of vegetation can be defined and enclosed, and the collecting nozzle used to suck up the invertebrates from it for a standard length of time. In general, apparatus with a large collecting nozzle (greater than 10–15 cm in diameter) is more suited to the first method, while those with a smaller nozzle are better suited to the second.

After the sample has been taken, the net bag containing the invertebrates should be sealed and placed in a killing bottle (p. 145) and its contents then removed and preserved. Since the net will inevitably also contain large amounts of plant debris sucked up with the invertebrates, the contents of the net bag will have to be sorted, ideally in water and/or by flotation (p. 160).

Advantages and disadvantages

Suction sampling is only effective in vegetation less than 15 cm high, which has not been flattened by wind, rain, or trampling. Like sweep netting, it cannot be used if the vegetation is damp. Suction sampling collects fewer invertebrates per unit time spent in the field than sweep netting does. However, although extraction efficiency varies to some extent in differently structured vegetation, this is usually less of a problem than it is with sweep netting. Hence suction sampling may often be the preferred option for monitoring invertebrates in low vegetation and sweep netting the preferred option for surveying them.

Suction samplers can be heavy and tiring to carry long distances and require refilling with a petrol/oil mix at frequent intervals. Being mechanical, they are also prone to breaking down. In particular, modified suction samplers like that shown in Figure 4.9 are susceptible to getting their carburettor clogged with debris from the petrol tank. It is therefore important to strain the fuel through a funnel with a fine filter on it when refilling the petrol tank. Suction samplers are also expensive.

Biases

Suction sampling under-records large invertebrates (> 3 mm long) that can take shelter (e.g. hunting spiders) or that are firmly attached to the vegetation, e.g. Lepidopteran larvae. They will also probably under-record species living low down in tall vegetation, and species that can take evasive action when they sense the noisy sampler approaching.

Pond nets and tow nets

Mainly free-swimming (nektonic) invertebrates and zooplankton and those resting on the surface of the substrate

Method

Pond nets can be used as a quick method of catching large numbers of aquatic invertebrates. When surveying invertebrates, it is a good idea to try a variety of techniques: moving the net in a figure of eight, just above the bottom of the water, so that invertebrates on the substrate (e.g. bugs) are stirred up and caught as they swim away, pressing the net rim against mossy stones to catch tightly clinging nymphs, and moving the net at different speeds and depths through open water and patches of aquatic vegetation. The net should be twisted at the end of its movement through the water, so that the net bag flips over the frame and so prevents its contents from escaping.

After taking the net out of the water, it should be allowed to drain and the net contents should be emptied onto a white tray (e.g. a photographic developing tray), taking care to wash all the specimens off (some nymphs cling very tightly). Specimens can be picked up using a small pipette and placed in temporary storage containers. It is important to wash the net carefully before moving on to the next site to prevent invertebrates inadvertently being recorded from the wrong site. Small tea strainers can be useful for deftly catching water beetles and bugs.

Netting can be standardised for use in comparing invertebrate numbers between sites or at the same site over time, either by taking a standard number of equal-length net hauls, or by sampling for a set period of time. An easy way to measure the length of net hauls is to mark two points in the water using canes, and then pull the net at as constant a speed as possible between these. In very soft substrates in water less than 2 m deep, hand nets can be used to take benthic samples. Sample sizes can be standardised by stirring up the benthos and then taking a subsample of known volume from it (e.g. Mason 1977).

Tow nets can be used from boats to produce quantitative estimates of zooplankton numbers in open water. A large number of designs have been used, that shown in Figure 3.5 being a typical example. Variations on this basic design include addition of fins to stabilise the net, a 'stretching bar' to keep the collecting bottle a set distance from the net frame to keep the net taught, and the addition of a flow meter to measure the distance over which the net has been pulled. It is important that the opening of the net is kept perpendicular to the water flow, both to reduce turbulence and maintain the same-size net opening throughout each tow.

Once a tow has been finished, the contents of the collecting bottle should be emptied into a container and all the zooplankton stuck to the inside of the net washed off by splashing water against the outside (never the inside) of the net. This can take a considerable time. Zooplankton are best sorted by then subsampling the contents of the container. This is most easily done by gently stirring up the water, so that invertebrates in it are as evenly dispersed as possible, and then removing known volumes of water. Invertebrates in these subsamples can then be counted and identified by pouring each subsample into a Petri dish on a black or white background, using a microscope if required.

Advantages and disadvantages

Pond nets are easy to use in any open water that is close to a bank or shallow enough to wade into. They are particularly useful for carrying out quick surveys of sites. For example, Furse *et al.* (1981) found that sampling of ponds using a hand-held net for just three minutes collected approximately 62% of families and 50% of species that could be found by 18 minutes of netting. For monitoring, the most serious disadvantage of the use of this method is the difficulty in standardising the way in which the net is pulled through the water.

Tow nets are only of practical use in relatively large areas of open, weed-free water. They need to be operated from a boat, and therefore usually require two people: one to control the boat while the other looks after the tow net. The speed of the boat can be difficult to control (see below). Nets can become clogged with sediment and phytoplankton and zooplankton, and this will reduce their catching efficiency. It can also be difficult to arrange the position of the net so that it is not subject to water turbulence caused by the boat. Tow nets have the advantage of sampling large areas of habitat, useful when sampling zooplankton which often have a patchy distrubution.

Biases

Catches will vary with the speed that the net is pulled through the water. Slow movement will fail to catch invertebrates that are able to avoid the approaching net. Fast movement will push forward the water in front of the net and again fail to catch a significant proportion of invertebrates in its path. The invertebrate fauna will vary with depth, and the depth at which the net is pulled through the water can be difficult to keep constant. Unless specifically dragged close to the bottom of the water, standard netting will underestimate numbers of invertebrates resting on the substrate (e.g. resting prawns, shrimps, and corixids).

Cylinder samplers
Small nektonic crustaceans and other zooplankton

Method

Cylinders can be used to delimit and remove columns of water of known diameter and depth, containing small crustaceans and other zooplankton. The cylinder should be lowered vertically or horizontally (in the case, for example, of sampling invertebrates in gaps between aquatic plants), quickly enough that invertebrates are unable to flee from the approaching cylinder, but not fast enough to cause unnecessary turbulence in the water. A bung should then be placed into the bottom of the cylinder (or both ends if used horizontally), the cylinder removed, and its contents emptied into temporary storage containers. Small invertebrates are best sorted by subsampling the water collected, as described on p. 168.

Advantages and disadvantages

Cylinders samplers are only practical for use in water shallow enough to wade into or areas very close to the water's edge, but unlike tow nets they can be used to sample amongst relatively weed-filled water. This method only samples relatively small volumes of water, and zooplankton often have a very patchy distribution. Hence it may be necessary to take a large number of samples.

Biases

Some zooplankton may move away and avoid capture because of disturbance caused during sampling. Others may actively avoid the approach of the cylinder. The speed at which the cylinder is moved through the water will influence the latter.

The Robertson dustbin sampler

Free-swimming (nektonic) invertebrates and those resting on the surface of the substrate

Method

The Robertson dustbin sampler (Figure 4.10) should be rapidly dropped through shallow water, with the net crumpled up in the middle of the bottom of the sampler. To avoid scaring off aquatic invertebrates, it is best to wade stealthily through the water, and then quickly drop the sampler at arm's length perpendicular to the direction of wading. The strings on the side of the sampler should then be pulled up, so that the net spreads across the bottom of the sampler. At this point the whole sampler can be carefully lifted out of the water and the number of invertebrates caught in the netting identified and counted.

Advantages and disadvantages

The Robertson dustbin sampler is quick to use and suitable for any open water with a relatively even bed, where it is possible to wade.

Biases

More active invertebrates, particularly shrimps and prawns, may take evasive action and therefore be under-recorded because of the disturbance caused by the approach to the observer, together with that caused by the sampler being lowered into the water.

Bait traps

Scavenging planarians, leeches, and crustaceans

Method

Traps baited with meat can be used to catch a variety of scavenging invertebrates. The bait (a slit earthworm, fresh liver, meat, tinned sardines, etc.) should be placed in a jam jar with one or more openings made in its lid. The diameter of the opening(s) can be used to determine what size of and hence which groups of invertebrates are caught (e.g. small holes will allow triclads in but exclude leeches and crayfish). The trap should be checked at regular intervals.

Advantages and disadvantages

This method is cheap and easy, but difficult to standarise for use in comparing numbers of invertebrates between sites or at the same site over time (see below).

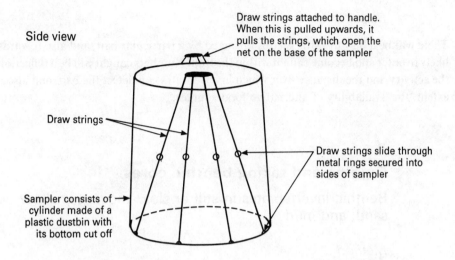

Side view

Draw strings attached to handle.
When this is pulled upwards, it
pulls the strings, which open the
net on the base of the sampler

Draw strings

Draw strings slide through
metal rings secured into
sides of sampler

Sampler consists of
cylinder made of a
plastic dustbin with
its bottom cut off

View of base of sampler with:

(a) net closed   (b) net partially open   (c) Net fully open

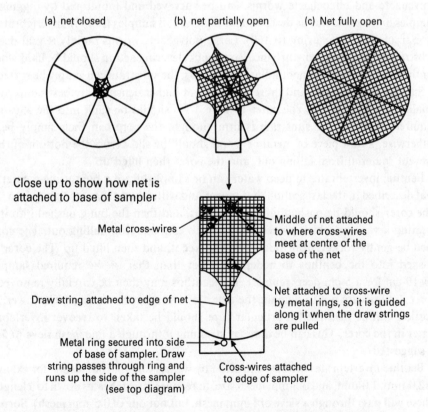

Close up to show how net is
attached to base of sampler

Metal cross-wires

Middle of net attached
to where cross-wires
meet at centre of the
base of the net

Net attached to cross-wire
by metal rings, so it is guided
along it when the draw strings
are pulled

Draw string attached to edge of net

Metal ring secured into side
of base of sampler. Draw
string passes through ring and
runs up the side of the sampler
(see top diagram)

Cross-wires attached
to edge of sampler

Figure 4.10 A dustbin sampler for catching free-swimming (nektonic) invertebrates
and those resting on the surface of the substrate (from Robertson 1993).

Biases

There will be a bias towards species attracted by a particular bait, and also towards species likely to enter and become caught within the trap. Numbers caught will be a reflection of both the activity and the distance over which invertebrates can detect the bait, and also, to some extent, the availability of alternative food sources.

Digging and taking benthic cores

Benthic invertebrates in still or slow-moving shallow water, sand, and mud

Method

On estuaries and sandy or muddy shores, large, low-density invertebrates such as various polychaete and oligochaete worms, can be surveyed and monitored by digging substrate samples in the same way as described for digging soil samples (p. 158). Invertebrates can then be extracted by wet sieving (p. 160). Large polychaete worms rapidly retreat deep into the substrate when sensing disturbance, so a quick, levering action should be used when digging for these. If surveying larger molluscs or worms, the substrate can simply be sorted by hand.

Smaller invertebrates and those occurring at higher densities are best sampled by taking smaller substrate cores. The corer (Figure 4.11) should be sunk into the substrate to the required depth. If the substrate is firm enough, the core can then simply be lifted up. Otherwise, a thin piece of metal or wood should be slid across the bottom of the corer to prevent material from falling out, and the corer then lifted up.

Benthic invertebrates in deep water can be sampled from a boat using a corer similar to that described in the last section, but longer and with a bung that fits tightly into its top end. The corer should be pushed into the benthos, and then the bung pushed into its top end, creating a vacuum which prevents the corer's contents from falling out. The corer should then be gently rocked from side to side to free it, and then lifted up. The corer should be pushed into the benthos to a depth greater than that of the required sample (e.g. to 10–15-cm for a 5-cm-deep sample). The benthos may then be carefully removed from the corer, taking care not to damage the delicate invertebrates within it. The lower, unwanted portion can be discarded. Particular care should be taken to recover invertebrates from water in the corer. This can be done by straining it through a fine mesh sieve (0.5-mm mesh is suggested).

Benthic invertebrates are best extracted by wet sieving (p. 160), using, for example, sieves of 2.0-mm, 1.0-mm, and 0.5-mm mesh size, to recover invertebrates down to a length of 1-mm (these will pass through a sieve of 1-mm mesh, but not one of 0.5-mm mesh). Sorting may be made quicker and more efficient by adding a 1% solution of Rose Bengal dye, which stains translucent invertebrates pink.

(a)

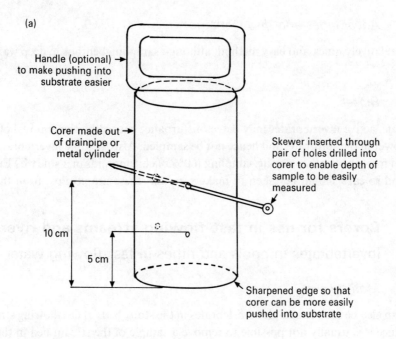

Handle (optional) → to make pushing into substrate easier

Corer made out → of drainpipe or metal cylinder

Skewer inserted through pair of holes drilled into corer to enable depth of sample to be easily measured

10 cm

5 cm

Sharpened edge so that corer can be more easily pushed into substrate

(b)

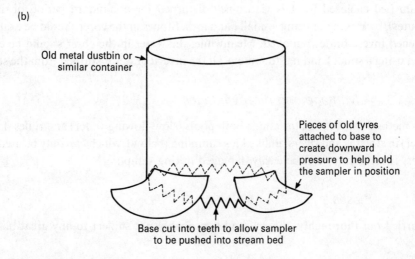

Old metal dustbin or → similar container

Pieces of old tyres attached to base to create downward pressure to help hold the sampler in position

Base cut into teeth to allow sampler to be pushed into stream bed

Figure 4.11 Corers. (a) is for use in soft substrates and (b) in pools and riffles in fast-flowing water.

This is a relatively quick and easy method, although sampling benthos in deep water can be difficult.

Some large, active invertebrates may detect disturbance caused by the removal of samples, retreat lower in the substrate, and hence not be sampled. Also, benthic invertebrates may be dislodged from the benthos during sampling (60% of chironomids in a study by Euliss *et al.* 1990), and so care should be taken to make sure that these are not lost from the corer.

Corers for use in fast-flowing streams and rivers

Invertebrates in pools and riffles in fast-flowing water

Method

Corers can also be used to sample invertebrates in the stony beds of fast-flowing streams and rivers. Since it is usually not possible to remove a sample of the stream bed in the corer, a slightly different method is used. A larger corer (Figure 4.11) is used to delimit an area of the stream bed. Once the corer has been sunk into the stream bed as far as it will go, the area of stream bed enclosed by it is vigorously disturbed for a standard period of time (10–15 minutes), by kicking or using a small hand fork. Stones in the corer should be examined, and attached invertebrates removed. Meanwhile, the water in the corer should be continually sifted using a small hand net to remove all the invertebrates disturbed from the stream bed.

Advantage and disadvantages

This method can be used to sample both pools (slow-flowing water) and riffles (fast-flowing water) in streams and rivers, unlike kick sampling (below), which can only be used to sample riffles. Invertebrates can be easily damaged during sampling.

Biases

If carried out thoroughly, this method is probably not subject to any great bias.

Kick sampling

Invertebrates in riffles in fast-flowing water

Method

The majority of invertebrates in fast-flowing streams and rivers are found amongst stones and gravel on the stream bed. Kick sampling involves dislodging invertebrates in the stream

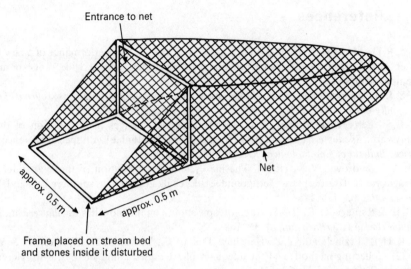

Entrance to net

Net

approx. 0.5 m

approx. 0.5 m

Frame placed on stream bed
and stones inside it disturbed

Figure 4.12 A Surber sampler for collecting invertebrates dislodged from the river bed in fast-flowing water.

bed by kicking and disturbing the substrate and catching the dislodged invertebrates in a net held a short distance downstream. The technique is easily standardised and widely used to obtain 'scores' of macroinvertebrates for use in water quality assessment. A widely used method is to kick-sample all the microhabitats along a stretch of river for a total of three minutes using a 0.9-mm-mesh pond net. Individual 'scores' for macroinvertebrate taxa, based on their relative tolerance to oxygen depletion, are then summed to produce a biotic score for that section of the river (Armitage *et al.* 1983). Invertebrates are most easily sorted in a white (or black and white) tray as described on p. 146.

A more refined method of kick sampling involves the use of a 'Surber sampler' (Figure 4.12). The same method is used, the area of bed to be sampled being defined by the frame resting on the substrate.

Advantages and disadvantages

This method is quick, but since only a proportion of invertebrates from the stream bed are caught, it is only a relative method, and cannot be used to estimate population densities.

Biases

Both methods of kick sampling will tend to under-record invertebrate species firmly attached to stones, and heavy species that are unlikely to be carried by the water and caught in the net, such as stone-cased caddis fly larvae.

References

Armitage, P. D., Moss, D., Wright, J. F. & Furze, M. T. (1983). The performance of a new biological water quality score system based on macro-invertebrates over a wide range of unpolluted running-water sites. *Water Research* **17**, 333–347.

Brooks, S. J. (1993). Review of a method to monitor adult dragonfly populations. *Journal of the British Dragonfly Society* **9(1)**, 1–4.

Brown, E. S., Betts, E. & Rainey, R. C. (1969). Seasonal changes in distribution of the African Armyworm, *Spodoptera exempta* (Wlk) (Lepidoptera: Noctuidae) with special reference to eastern Africa. *Bulletin of Entomological Research* **58**, 661–728.

Cherrill, A. J. & Brown, V. K. (1990). The life cycle and distribution of the Wart-biter (*Decticus verrucivorus* (L.) (Orthoptera: Tettigoniidae) in a chalk grassland in southern England. *Biological Conservation* **53**, 125–143.

Cragg, J. B. & Ramage, G. R. (1945). Chemotropic studies on the blowflies *Lucilia sericata* (Mg.) and *Lucilia caesae* (L.). *Parasitology* **36**, 168–175.

Disney, R. H. L., Erzinclioglu, Y. Z., Henshaw, D. J. de C., Unwin, D. M., Withers, P. & Woods, A. (1982). Collecting methods and the adequacy of attempted fauna surveys, with reference to the Diptera. *Field Studies* **5**, 607–621.

Disney, R. H. L. (1986). Assessments using invertebrates: posing the problem. In *Wildlife Conservation Evaluation* ed. by M. B. Ussher, pp. 271–293. Chapman & Hall, London.

Disney, R. H. L. (1987). Rapid surveys of arthropods and the ranking of sites in terms of conservation value. In *Biological Surveys of Estuaries and Coasts*, ed. by J. M. Baker & W. J. Wolff, pp. 73–75. Estuarine and Brackish-Water Sciences Foundation Handbook, Cambridge University Press, Cambridge.

Ditlhogo, M. K. M., James, R., Laurence, B. R., & Sutherland, W. J. (1992). The effects of conservation management of reed beds. 1. The invertebrates. *Journal of Applied Ecology* **29**, 265–276.

Euliss, N. H., Swanson, G. A. & Mackay, J. (1992). Multiple tube sampler for benthic and pelagic invertebrates in shallow wetlands. *Journal of Wildlife Management* **56(1)**, 186–191.

Furse, M. T., Wright, J. F., Armitage, P. D. & Moss, D. (1981). An appraisal of pond net samples for biological monitoring of lotic macroinvertebrates. *Water Research* **15**, 679–689.

Greenslade, P. J. M. (1964). Pitfall trapping as a method for studying populations of carabidae. *Journal of Animal Ecology* **33**, 301–310.

Greenslade, P. & Greenslade, P. J. M. (1971). The use of baits and preservatives in pitfall traps. *Journal of the Australian Entomological Society* **10**, 253–260.

Hill, C. J. Gillison, A. N., & Jones, R. E. (1992). The spatial distribution of rain forest butterflies at three sites in north Queensland, Australia. *Journal of Tropical Ecology* **8**, 37–46.

Lasiak, T. (1991). The susceptibility and resilience of rocky littoral molluscs to stock depletion by the indigenous coastal people of Transkei, Southern Africa. *Biological Conservation* **56**, 245–264.

Luff, M. L. (1975). Some factors influencing the efficiency of pitfall traps. *Oecologia* **19**, 345–357.

Luff, M. L. (1986). Aggregation of some Carabidae in pitfall traps. In *Carabid Beetles: Their Adaptations and Dynamics*, ed. by P. J. den Boer, M. L. Luff, D. Mossakowski & F. Weber. Gustav Fischer, Stuttgart & New York.

Mason, C. F. (1977). Populations and production of benthic animals in two contrasting shallow lakes in Norfolk. *Journal of Animal Ecology* **46**, 147–72.

Moore, N. W. (1964). Intra- and interspecific competition among dragonflies (Odonata). *Journal of Animal Ecology* **33**, 49–71.

Moore, N. W. & Corbet, P. S. (1990). Guidelines for monitoring dragonfly populations. *Journal of the British Dragonfly Society* **6(2)**, 21–23.

Morgan, N. C. (1971). Factors in the design and selection of insect emergence traps. In *A Manual on Methods for the Assessment of Secondary Productivity in Fresh Waters*, ed. by W. T. Edmondson & G. G. Winberg. IBP Handbook No. 17, Blackwell Scientific Publications, Oxford.

Mundie, J. H. (1971). Techniques for sampling emerging aquatic insects. In *A Manual on Methods for the Assessment of Secondary Productivity in Fresh Waters*, ed. by W. T. Edmundson & G. G. Winberg. IBP Handbook No. 17, Blackwell Scientific Publications, Oxford.

Murphy, P. W. (1962). Extraction methods for soil animals. In *Progress in Soil Zoology*, ed. by P. W. Murphy, pp. 75–114. Butterworths, London.

Nordstrom, S. & Rundgren, S. (1972). Methods of sampling lumbricids. *Oikos* **3**, 344–352.

Pollard, E. (1977). A method for assessing changes in the abundance of butterflies. *Biological Conservation* **12**, 115–134.

Pollard, E. (1979). A national scheme for monitoring the abundance of butterflies: the first three years. *British Entomological and Natural History Society, Proceedings and Transactions* **12**, 77–90.

Pond Action. (1989). *National Pond Survey. Methods Booklet*. Pond Action, Oxford Polytechnic, Oxford.

Raw, F. (1960). Earthworm population studies: a comparison of sampling methods. *Nature, Lond.* **187**, 257.

Robertson, P. A. (1993). The management of artificial coastal lagoons in relation to invertebrates and avocets *Recurvirostra avosetta* (L.). Ph.D. thesis, University of East Anglia, UK.

Stubbs, A. & Chandler, P. (1978). *A Dipterist's Handbook. The Amateur Entomologist* Vol. 15. The Amateur Entomologist's Society, Harmondsworth.

Thomas, J. A. (1983). A quick method for estimating butterfly numbers during surveys. *Biological Conservation* **27**, 195–211.

Thomas, J. A. & Simcox, D. J. (1982). A quick method for estimating larval populations of *Melitaea cinxia* during surveys. *Biological Conservation* **22**, 315–322.

Usher, M. B. (1990). Assessment of conservation values: the use of water traps to assess the arthropod communities of heather moorland. *Biological Conservation* **53**, 191–198.

Volkl, W., Zwolfer, H., Romstock-Volkl & Schmelzer, C. (1993). Habitat management in calcareous grasslands: effects on the insect community developing in flower heads of Cynarea. *Journal of Applied Ecology* **30**, 307–315.

Wright, A. F. & Stuart, A. J. A. (1992). A study of the efficacy of a new inexpensive type of suction apparatus in quantitative sampling of grassland invertebrate populations. *British Ecological Society Bulletin* **23**, 116–20.

5 Fish

Martin R. Perrow

ECON Ecological Consultancy, School of Biological Sciences, University of
East Anglia, Norwich NR4 7TJ, United Kingdom

Isabelle M. Côté

School of Biological Sciences, University of East Anglia, Norwich NR4 7TJ,
United Kingdom

Michael Evans

National Rivers Authority, Anglian Region, 79 Thorpe Road, Norwich NR1 1EW,
United Kingdom

Three-quarters of the world's surface area is covered by water.
Three-quarters of the world belongs to the fish.
(Attenborough 1979)

Introduction

Fish are the most abundant, widespread, and diverse group of vertebrates, comprising 22 000
species with a dazzling variety of form, size, and habits. To exploit this profusion of potential
food, humans have devised a fantastic array of gears to capture fish. Ecologists rely heavily
on modified forms of these methods to census fish populations, since fish are often difficult to
observe in their natural habitat.

Sampling fish requires a high level of resources (e.g. time, labour, cost of equipment), and
this increases with the size of the habitat (e.g. a pond versus the sea). Many commercial-scale
techniques (e.g. deep-sea seines and trawls) are beyond the scope of ecologists, but they can
be scaled down to suit smaller habitats. To census fish in the largest aquatic systems, we
recommend using the data from commercial catches, where available, or visiting markets
where fish are landed. The use of local knowledge and technology, particularly where
resources are limited, is always recommended.

Methods of capturing fish fall into two categories: *passive* methods, which rely on the fish
swimming into a net or a trap, and *active* methods, where fish are pursued. The fact that fish
are poikilothermic influences the choice of method and timing of sampling. For example, in
temperate zones, active methods may be more successful in winter when fish are less mobile,
whereas passive techniques may work best in summer when fish are more active. The choice
of method will also be guided by gear selectivity. Most techniques are selective, and the
limitations of many methods, even those in widespread use, are unresolved. There are few
truly quantitative techniques, so statistics such as catch per unit effort (CPUE) are
commonly used to generate indices of abundance (Cowx 1991). Selectivity stems from the
physical features of the gear, e.g. mesh size, but it is also influenced by the ecology of the
target species. Variability in swimming speed, seasonality as a result of migration, diurnality,

178

Table 5.1. *Fish censusing techniques and their suitability in various habitats*

★ usually applicable, +often applicable, ? sometimes applicable. The page number for each method is given.

Method	Shallow	Deep	Still	Slow flow	Fast flow	Fresh water/ brackish	Saline	Open	Vegetated	Coral reef	Page no.
Bankside counts	★		★	★		★		★			180
Underwater observations	+	★	★	★	?	★	★	★	?	★	181
Electrofishing	★		★	★	+	★		★	★		182
Seine netting	★		★	★		★	★	★			185
Trawling		★	★	★		★	★	★			188
Lift and throw nets	★	+	★	+		★	★	★	+	+	191
Push nets	★		★	★		★	★	★			191
Hook and line	★	★	★	★	★	★	★	★		★	192
Gill netting	★	★	★	+		★	★	★	+	+	194
Traps	★	★	★	★		★	★	★	+	★	196
Hydroacoustics		★	★	★		★	★	★	+		199

Table 5.2. *Fish-egg censusing techniques*

★usually applicable, + often applicable, ? sometimes applicable. The page number for each method is given.

	Benthic eggs in nest	Pelagic eggs	Page no.
Visual estimate	★		200
Volumetric estimate	★		201
Plankton nets		★	201
Emergence traps	★		202

and patchiness of distribution all influence how catchable a species is. It is therefore vital to have at least some knowledge of the ecology and behaviour of fish in the habitat to be sampled.

Finally, the selection of the technique will depend on the habitat to be sampled. Factors such as depth, clarity, presence of vegetation or speed of the current will need to be considered. A hydrographical survey prior to sampling may therefore be necessary.

This chapter provides only a general outline of the techniques which we consider most likely to be used in the course of ecological censusing of fish. We strongly recommend Nielsen & Johnson (1983) for more detailed discussion of these and other methods. Methods such as poisons (e.g. rotenone) and explosives have been used to census fish; however, we consider them too destructive for general use, and they are not included here. Most methods are aimed at censusing juvenile and adult fish, but many can be modified to sample larval fish as well.

Several methods also exist for counting fish eggs. A few are covered in this chapter, but Smith & Richardson (1977) provide a good review of standard techniques.

Bankside counts

Conspicuous fish in pools and slow-moving, shallow, freshwater streams and small rivers

Method

It is sometimes possible to census fish without catching them and without getting wet. Bankside counts are a good technique in shallow, slow-moving, and clear waters with minimal vegetation, such as streams or even lake shores.

The stretch of water to be surveyed is usually divided into contiguous but non-overlapping sections. The sections should be small enough that all fish can be counted from a single vantage point. The use of landmarks on the shore helps in delimiting the sections. Observers, wearing polarising sunglasses, to reduce glare, and dull coloured clothes, should move slowly to the vantage point and conceal themselves behind riparian vegetation. Artificial structures, such as docks or bridges, may be used when available. Once in position, wait motionless for at least five minutes before counting to minimise the effects of disturbance. Counts are best made on sunny days or at least bright overcast days. Intensive sunshine can create problems with glare, and shadows can betray the presence of the observer, whilst rain, wind, and surface ripples make observations nearly impossible. After the count, move away from the bank before proceeding to the next observation site. Fish density can be calculated by measuring the area surveyed.

Advantages and disadvantages

Shore-based visual counts are cheap, fast, and easy. They are particularly appropriate when the water is too shallow to be sampled easily by other techniques and are especially useful for censusing particular age-classes of fish (e.g. young-of-the-year) which seek shallow-water habitats.

Direct observation (see also Underwater observations, p. 181) allows parameters such as fish size and sex and many features of the environment (e.g. water depth, current speed, substrate type, vegetation cover) to be recorded. It also allows detailed behavioural observations to be made. Stress to the fish is minimal.

One potential problem is that fish are very aware of human presence, and any disturbance will reduce the accuracy of visual estimates. Fish can take a long time to leave cover and resume their activities following disturbance. Visual counts from banks typically have high inter-observer variability (Hankin & Reeves 1988), although having a few, well-trained people counting fish can reduce this problem.

Biases

Big and brightly coloured fish are easier to see than smaller, duller-coloured fish. This may introduce a sex-bias since males are often larger and/or more brightly coloured than females. Age-related behaviour may also introduce a bias if some age-classes seek cover more than others. Fish in deeper water, turbulent areas, turbid waters, areas of high surface glare, or dense habitat may be overlooked.

Underwater observations

Fish in clear, calm, shallow marine (especially coral reefs) or fresh water

Method

There are two ways of gathering underwater observations of fish: snorkelling and SCUBA diving. The choice depends mainly on the clarity of the water and the depth at which observations must be made. Observations made deeper than 1–1.5 m in turbid freshwater lakes or 3–4 m in clear tropical waters will generally require SCUBA.

With both techniques, transect or point-count sampling may be used. For transects (see Chapter 2, p. 54), the observer swims along a fixed rope or chain, marked at 1–5-m intervals, laid on the substrate. The numbers and sizes of fish of each species occurring within a given distance of the transect line are recorded. In clear tropical waters, this can be up to 5 m on either side of the transect, whereas in temperate waters, the lower visibility may reduce the transect width to 1–2 m. It is best to record fish that are somewhat ahead of the observer (the distance will depend on the visibility), as some species will leave the transect corridor if the observer comes too close. Swimming speed is important when running transects because the faster you swim, the less accurate the survey becomes. Swimming speed should therefore be slow and constant.

Data are most easily recorded by writing with a soft lead pencil on a perspex sheet from which the glaze has been removed by abrasion (sandpaper or steel wool works well). The writing may later be rubbed off with an eraser.

Point-count sampling (see Chapter 2) is undertaken from a given location underwater and is generally done using SCUBA. It is best for the diver to sit on the bottom for several minutes before starting the count, to let the fish habituate to the diver's presence. After the count, it is necessary to estimate the radius of visibility in order to calculate the area surveyed.

Both methods give good measures of abundance, density, and species diversity. Calibration of size estimates made underwater may be necessary and will involve catching fish that have been measured visually. Remember that things look about 30% larger underwater than they really are.

Advantages and disadvantages

Underwater observations are a favoured way of censusing fish because of the ecological insights gained by sharing the fish's environment. Underwater observations are not destructive and cause minimal disturbance. It is relatively easy to train people to dive or snorkel, but the reliability of underwater censuses depends on accurate species identification. This can be challenging, particularly in the tropics, where species diversity can be great. While underwater, several other facets of the environment and behaviour can also be recorded.

Snorkelling equipment is cheap compared to SCUBA gear. Although there are inherent risks to both techniques (e.g. drowning or getting eaten by a shark), there are more safety concerns with SCUBA. SCUBA requires proper certification, the presence of a diving partner, nearby tank refilling facilities, and a decompression chamber within reach. While there is no time limit to snorkelling or diving near the surface, there are strict rules about the time that can safely be spent deeper than 10 m. The deeper the dive, the shorter the permitted bottom time will be, which will limit the length of surveys. Getting wet invariably means getting cold, even in tropical waters, although the use of dry suits may help to alleviate this problem.

Biases

Conspicuous species, ages, and sexes are more likely to be recorded. The timing of the census is important; on coral reefs, for example, the composition of fish communities changes markedly over the course of the day or night. In heavily fished areas, especially where spearfishing takes place, larger food fish may avoid the observer.

Electrofishing

All fish in shallow, relatively clear, fresh or brackish waters, particularly vegetated areas

Method

Electrofishing involves passing an electric current through water via electrodes (anode and cathode) which stuns nearby fish, leading to their disorientation and easy capture. Power is supplied by an electrical generator (or batteries in the case of backpack units) and is converted to the required form by an electrofishing unit or box. The circuit is completed by on/off switches on the anodes. Several current types may be used, each producing slightly different effects. The most commonly used is DC, because it attracts fish to the anode and causes fewer harmful effects to the fish than AC. However, because DC current requires large generators and electrofishing units, pulsed DC is often used instead.

During electrofishing, anodes are often hand held, whilst the cathode (a multistrand copper wire or even a metallic plate) trails behind the boat or operator. The charge is usually kept on during fishing. The *key* to electrofishing is always to be close enough to the target fish to induce a response and to 'explore' all available habitat with the anodes. In addition, the operator should always work in an upstream direction as disturbed sediments then flow away from the sampling area and stunned fish drift toward the operator. In narrow (relative to the span of the electrodes) streams and rivers, fish are captured efficiently (Kennedy & Strange 1981) and absolute measures of abundance may be generated.

In small streams, the operator typically wades in the water wearing rubber waders, with the battery-powered gear on his/her back, holding the anode in one hand and a dip net in the other. If greater power is required, generators and boxes may be left on the bank, with long leads connecting them to the anode. Stop-nets can be stretched across the stream both to delineate the area to be sampled and to prevent fish from swimming out of the area. Fishing operators then systematically and continuously search the stream with the anodes moving upstream from the downstream stop-net. Stunned fish are removed with dip nets and placed in containers of water to recover. This usually takes a few seconds, but may take several minutes for large fish. On warm days or when many fish are encountered, the water in the containers should be aerated. The whole fishing procedure is then repeated a fixed number of times (usually three) at the same site. This type of 'depletion fishing' allows an absolute estimation of fish population size and/or density (see Chapter 2, p. 40).

In deeper streams or canals, electrofishing can be undertaken from a non-metallic boat carrying both staff and gear. The operator is usually at the bow holding the anode ahead of the boat, while a crew steers with oars or engines. The boat may also be pulled on ropes by operators on the banks. If the river is much wider than the boat, a tight zig-zag pattern of search can be adopted which allows a more thorough sampling of all habitats. Alternatively, the river may be divided into manageable sections with stop-nets, more than one team may operate in parallel, or multiple anode systems may be mounted on booms (see Cowx *et al.* 1990).

In very large rivers and lakes, transects or point sampling (Chapter 2) are preferred. Both may be random or systematic. During transects, the power is usually kept on as the boat cruises across a predetermined route, and stunned fish are collected as they encounter the electric field. In point sampling, the power is only supplied at predetermined points as the boat moves across the water body. Point sampling is particularly effective in vegetated areas and for larval and small fish. Good estimates of density and biomass may be obtained by calculating the area sampled at each of a large number of points (Persat & Copp 1990).

Frightening fish from the area to be sampled is a potential problem which can be overcome in part by using oversized anodes to increase the effective sampling zone or by altering the method of propulsion (e.g. oars versus engine). Depending on the circumstances, it may pay to move either quietly and slowly or extremely rapidly so fish have little time to react.

In large systems, relative measures of abundance, such as catch per unit effort (CPUE), are more appropriate than absolute estimates. CPUE is also valuable in monitoring population change over time if time of fishing and area covered are standardised.

Several environmental factors, such as water conductivity, influence the efficiency of electrofishing, and there are various ways to adjust the gear to suit particular environmental conditions (Box 5.1). The capture of stunned fish is greatly affected by water clarity. Fish cannot be sampled in turbid water, and even if clarity is not limiting, experience suggests that the maximum depth for efficient fish capture is around 2 m. Vision (and concentration) is dramatically improved by the use of polarising glasses and large hats to reduce glare. Water temperature affects fish swimming speed and likelihood of escape. Each group of fish appears to have an optimal temperature range for capture. Salmonids, for example, are best caught between 5 and 10°C, and temperate cyprinids, between 10 and 20°C (Zalewski & Cowx 1990).

Box 5.1. **The effect of water conductivity on electrofishing, and how to adjust gear to maximise stunning efficiency**

- As conductivity decreases, there is a corresponding decrease in electrode current and a reduction in the stunning radius.
  Solution: Increase voltage output and/or increase anode size.
- At very low conductivity (<50 microsiemens), the fish is more conductive than the water when close to the anode, resulting in potential injury.
  Solution: Use a higher-frequency system.
- At very high water conductivity (>1000 microsiemens), current density is greater with an increased potential gradient along the length of the fish, giving good, low-threshold responses and thus requiring lower voltage. However, wastage of current between the electrodes is high.
  Solution: Increase current supply to the electrodes (i.e. use a bigger generator and units) or alternatively, reduce energy requirements by decreasing the size of the electrodes (at the cost of reduced stunning efficiency).

Advantages and disadvantages

Mixing water and electricity is inherently dangerous, and people have been killed while electrofishing. This technique therefore requires a high level of training of the operators and rigorous safety standards. In many countries, only licensed operators may electrofish. Basic safety procedures include wearing rubber gloves and boots at all times and avoiding immersion of any unprotected body parts. Equipment should have automatic dead-man switches on the anodes, immersion switches in backpack units, and emergency stop buttons, all of which are designed to break the circuit immediately if any of the operators fall in the water. For safety reasons a relatively high number of personnel are required (minimum of three in the UK). The cost of quality equipment is high and regular maintenance of gear is not only critical to ensure high-quality performance, but may be required by law. Such maintenance may be difficult in remote areas.

Electrofishing may injure or kill fish and other aquatic wildlife (including mammals, birds and amphibians) if they are exposed to current for too long. Injury to fish occurs most often when large numbers are encountered simultaneously. Operators should always be prepared to turn off the gear and collect all stunned fish before resuming fishing.

Electrofishing is a flexible technique that may be adapted for use in most freshwater and brackish water bodies. It disturbs the habitat less than other active methods and does not subject fish to abrasion (as in netting and trawling). Its main advantage over most other methods is that it draws fish from cover, making it the method of choice for sampling in vegetated waters.

Finally, to be performed to its full potential, electrofishing requires high motivation, concentration, visual acuity and good hand–eye coordination, physical endurance, and the ability to work in a team.

Biases

Large fish respond to electrofishing better than small fish, and long, thin fish are stunned more efficiently than short, stocky fish of equivalent weight (Zalewski 1983). Benthic species and those preferring dense cover may not rise to the surface, making their retrieval difficult. Territorial ones may attempt to maintain their position when approached and are thus relatively easy to capture. By contrast, shoaling species tend to flee, which makes them difficult to catch, but if they are encountered, many individuals may be captured simultaneously. Large fish and brightly coloured fish are easier to capture than small, cryptic fish. Operators may vary considerably in their ability to stun and collect fish.

Seine netting

Pelagic and/or demersal fish in open, still, or slowly flowing water

Method

A seine net is a wall of net fitted with floats at the top (float line) and a weighted line (lead line) on the bottom, and generally with a bulging section (bunt or bag) at the back of the net to hold the catch (Figure 5.1a).

The first step in seine netting is to encircle a known area of water. To do this, the net is generally fixed at one end, which can be the shore, a boat, or a buoy, and the rest of the net is laid out from a boat, or by walking, when in very shallow water, in an arc or semi-circular fashion and returning to the fixed point (Figure 5.1b). In shallow water, seine nets can also be dragged through the water by two people, one at each end, towards a fixed point (a boat or the shore, Figure 5.1b). A major potential problem is that the fish may be frightened away from the area while the net is being set. One way to alleviate this is to set the first part of the net as *quietly* as possible and then to close the gap to the fixed station as *quickly* as possible.

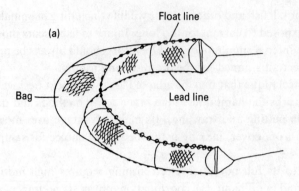

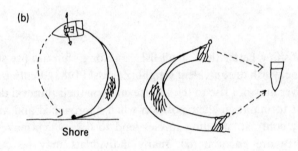

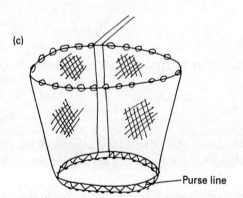

Figure 5.1   (a) Seine net. (b) Various methods of deploying a seine net: from a boat using a fixed point on shore (left), and in shallow water with two people dragging the net towards a boat or shore (right). (c) Purse seine.

The second step, hauling, is usually done by at least two people, each pulling on one end of the net (the number of people will depend on net length and drag). If the seine spans the entire water column (i.e. in shallow water), only the float line is pulled at first, to ensure that the lead line remains in contact with the bottom. In the closing stages of the haul, the lead line is pulled ahead of the float line and out from underneath the net, trapping the fish in the bag. If the seine spans only the top portion of the water column (i.e. in deeper water), both ends of the net must be hauled swiftly and simultaneously until only the bag remains in the water. During hauling, swimmers or boats may be used to frighten the fish into the bag. This is especially useful where fish try to jump over the float line. If the net is hauled too slowly, fish may escape from the mouth of the net. If it is hauled too quickly, the lead line may lift off the bottom or the float line may sink, again allowing fish to escape. This occurs particularly where fine sediments or macrophytes increase water resistance.

If a snag is encountered during hauling, it is best to cease hauling and send out a snorkeller, diver, or boat to investigate. Pulling from a different angle, particularly vertically, may be effective, and hauling may resume. However, if this is unsuccessful, it is best to abandon the haul and pull the net back into the boat, since damaging a net can be inconvenient and costly.

The final step, removing the catch from the net, is best achieved in small stages: removing fish with dip nets, pulling the net in a little closer, removing fish with dip nets, etc. Where large amounts of sediment are encountered, sluicing water through the net or rocking the net back and forth in the water may be effective, although great care must be taken not to damage the fish. Once most of the fish are removed, the net may be pulled onto the bank or boat. The net should then be cleaned and stacked neatly like an accordion (floats and leads separated) in preparation for the next haul.

Seines come in a huge range of lengths and depths. When one is sampling in shallow systems, the net should be 1.5 times the depth of water (Buckley 1987). This tends to prevent the net from lifting off the bottom during hauling. In deeper water, a purse seine (Figure 5.1c) may be more appropriate to catch fish in midwater and surface waters. Purse seines have rings attached to the lead line, through which a drawstring (purse line) passes. Once the net is set, the purse line is pulled and the bottom of the net is gathered together and upwards, thus trapping the fish.

The size of the net should be related to the size of the water body. In large systems, use the largest net that can be managed, but remember that the larger the net, the greater the effort required to haul it. In ponds and lakes, two people may be able to pull a 50-m-long net, but a 200-m-long one may require between 6 and 8 people. Large nets may be hauled mechanically, but hand pulling is usually preferred as this provides a better 'feel' for how the net is performing, the nature of the bottom, and whether any obstructions have been encountered.

Make as many hauls as possible to give a reasonable number of samples.

Advantages and disadvantages

Seine netting can be adapted to suit many situations, and as long as the net is small, it is efficient in terms of cost and labour. With large nets, the method quickly becomes expensive and labour-intensive. Furthermore, nets should be designed for specific water bodies (known depth of water, length required, specific mesh sizes to avoid sediments, etc.); hence a truly general-purpose net does not exist. One way of making a seine more versatile is to have detachable sections, perhaps with different mesh sizes, so that a net of the required length can then be 'made up' by tying particular sections together.

Seine nets can only be used in waters which are free of natural (trees, rocks, dense macrophytes, etc.) or artificial (wrecks, litter, and other human rubbish) obstructions. A hydrographical survey before sampling is advisable. Seining can cause great disturbance to the habitat sampled when macrophytes and sediment are swept in the net. Moreover, nets can damage and abrade fish, especially where sediment and gravels are retained in the net. Fine sediments can quickly deoxygenate the water and fish can suffocate while waiting to be removed in the final stages of the haul. Repairing damaged seine nets (and they will get damaged!) is a skilled operation, although small repairs can be done quickly with nylon rope or cable ties.

Biases

This is a relatively unselective technique with few biases, although fish in the littoral margins are not sampled adequately. There may be a tendency to miss fast-moving species during setting of the net and fish that seek refuge in the substrate during hauling.

Trawling
Slow-moving demersal and bottom-living species in large rivers, lakes and the sea

Method

Trawling involves towing a cone-shaped net along the bottom or through the water column at a specific depth. For midwater or surface trawls, the mouth of the net is fitted with floats at the top (headrope) and leads at the bottom (groundrope) to keep it open (Figure 5.2a). The simplest bottom trawl, the 'beam trawl' (Figure 5.2b), has a horizontal bar (beam) on the headrope, and sometimes a rectangular frame hanging from it helping to keep the mouth of the net open, and rows of chain (tickler chains) set in front of the groundrope to disturb fish buried in the sediments. Another common bottom trawl is the 'otter trawl', in which the mouth of the towed net is held open by water pressure against boards attached at an angle to the towing lines or warps (Figure 5.2c). Bottom trawls are sometimes mounted on sled runners which allow them to glide more smoothly over the substrate.

For both bottom and midwater trawls, the net is deployed either from the stern or from the side of a boat. It is essential that the deck area be clear and that the net be folded properly prior to fishing. In particular, make sure the warps are neatly coiled in storage containers (e.g. plastic dustbins). Cruising slowly, the cod end, marked with a buoy, should be introduced

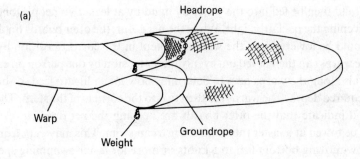

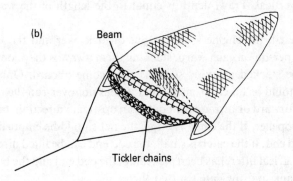

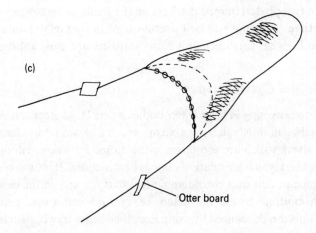

Figure 5.2 (a) Midwater trawl net. (b) Beam trawl (for bottom trawling). (c) Otter trawl (for bottom trawling).

first. The net should then be fed into the water by hand by at least two people, one on each side of the net, keeping the net fairly tight. When the net is out, the otter boards (in the case of otter trawls) should be lowered into the water and kept in the correct position by keeping warps tense. The warps can then be fed out evenly under tension by one person on each warp. Marks at intervals along the warps help to determine how much line to feed so that the net ends up at the required depth. The warps are then tied to the corners of the stern. The angle of the warps should indicate that the otter boards are keeping the net open.

The net must be towed at a faster rate than the fish can swim. This may vary from around 1.5 knots for slow-moving bottom fish to 5 knots or more for faster-swimming species. The length of the warps for bottom trawls should be approximately three times the depth of the water (Pereyra 1963). For midwater trawls, warp length will obviously depend on the depth of sampling. Towing a midwater trawl at a specific depth is a trade-off between towing speed and the sinking speed of the net. A quick way to determine trawl depth is to measure the angle ϕ of the warps relative to vertical. Trawl depth is equal to the length of the warps times $\sin(90°-\phi)$.

Once the trawl is completed, the engine is cut or put on tick over and the warps are retrieved by hand, with one person on each warp, stacking them away as they come in. The otter boards should also be stacked away as soon as they come aboard. Once the net approaches the boat, care should be taken to ensure that it does not over-run the boat. This may mean moving the boat forward or turning the boat in an upstream direction, taking care not to foul the net on the propeller. If the catch is large, the net should be emptied by hand nets before retrieving the cod end. If the catch is small, the cod end can be lifted directly onto the deck. If lots of detritus, e.g. leaf litter, has been picked up, the cod end and the body of the net are best towed to the bank and the catch sorted there.

To estimate the volume of water (for midwater trawls) or area of bottom (for bottom trawls) sampled, the distance covered by the net must be known. This can be determined either by towing between two predetermined markers on the bank, or by towing at a given speed for a set period of time. Trawling is at best a semi-quantitative method for estimating numbers and biomass; otherwise, catch per unit effort statistics are most appropriate.

Advantages and disadvantages

Trawling lends itself well to censusing in large water bodies where large areas can be covered in a short time. It is valuable on slowly flowing systems where flow would hinder the use of seine nets, and on tidal rivers which are too large or too saline for electrofishing.

Trawling is limited to waters which are relatively free of obstructions. It is time-consuming, requires expensive equipment, and uses considerable fuel (particularly in the case of otter trawls). The chains on bottom trawls are often heavy and can cause considerable environmental damage. This can be reduced by using electrified beam trawls. Fish tend to get damaged in the net, through crowding and abrasion.

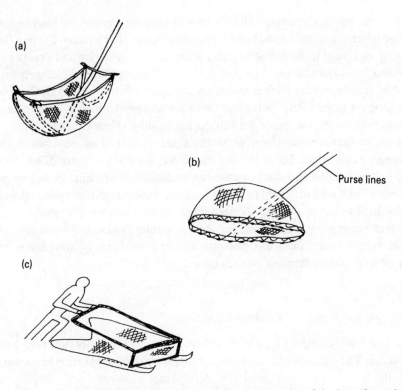

Figure 5.3 (a) Lift net (this particular design was used to fish elvers from the lower River Severn, UK). (b) Cast net with purse line. (c) Push net.

Biases

Slow-moving fish and those with poor escape responses are more likely to be caught, particularly in midwater trawls. Bottom-living species are sampled most reliably.

Lift, throw, and push netting

Small fish in shallow (or surface), still or slowly flowing water

Methods

Lift nets fall into two basic types: hand-held scoop nets (Figure 5.3a), which are simply inserted below the water surface and brought up sharply, and buoyant nets, which are allowed to lie on the bottom for a set period before coming up abruptly to the surface. As their name implies, buoyant nets are naturally positively buoyant. This is temporarily counteracted by attaching the net to a heavy frame or weight by means of dissolving tape or mint sweets with a hollow centre ('polo mints' in the UK). Once the attachment dissolves the net pops to the surface, catching fish on the way.

Cast nets are circles of netting, with weights around the perimeter. They usually have a central line which is retained in the hand for hauling the net after casting. Casting the net from the bank or especially from a boat requires great manual dexterity and practice to achieve distance and the correct shape of the net in the air to maximise the area sampled. Fish are captured by tangling in the meshes as the net collapses when it is hauled. One particularly efficient design is fitted with a purse line (as in purse seines) which is used to draw the net together to prevent the escape of the fish during hauling (Figure 5.3b).

Push nets are similar to trawl nets in having a pocket-shaped net attached to a triangular or D-shaped frame which keeps the mouth of the net open (Figure 5.3c). The frame is attached to long handles and can be pushed by wading or from a boat in shallow water. The frame is often fitted with rollers to facilitate pushing. Push-netting sampling effort can easily be standardised by always taking the same number of paces while pushing.

These techniques are typically used for small, shoaling fish in shallow waters, or near the surface of deeper water. They can generate absolute measures of abundance or biomass through discrete point sampling (see Chapter 2, p. 75).

Advantages and disadvantages

Lift, cast, and push netting are simple to perform and easily repeatable after one has acquired some practice. The equipment is fairly cheap, particularly if made with local materials. All these methods are used extensively in developing countries, and tapping into local knowledge will be a great advantage.

With a large number of samples, the methods may become labour-intensive. Lift and cast netting may be usable in highly vegetated habitats which are difficult to sample by other methods. Push netting is limited to sparsely vegetated, hard-bottomed habitats. A major disadvantage of lift and cast nets is that little is known of their efficiency and selectivity.

Biases

As the area sampled by lift, cast, and push nets is small, only small fish are likely to be captured effectively. Fast-swimming fish are more likely to escape.

Hook and lining

Large, predatory fish at low density, especially in fast-flowing rivers, deep lakes, and the sea

Method

This technique relies on catching fish with a baited hook attached to a line. For predatory fish the bait may be live invertebrates, such as worms, or pieces of meat, fish, etc. Artificial lures which imitate worms, flies, fish, frogs, and mammals may also be used. Bread, fruit, seeds, etc. may be used for other groups.

The line may be short and simply attached to the hand, or longer and strung to a pole. Rods with reels are also often used. Using poles and rods means that the fish do not have to be approached so closely and are less likely to be disturbed. They also give more control of the hook. The line with baited hook is then 'cast', from shore or from a boat. The line can be periodically pulled to mimic the movement of live bait. Many hooks may be used on a line to increase the chances of catching fish.

When a fish bites at the hook, the line becomes tense and should be pulled up immediately. A float attached to the line can give a clear indication that a fish is biting since it will sink when a fish pulls on the hook. Rapid upwards action then ensures that the hook becomes embedded in the fish's mouth. As soon as the fish nears the surface of the water, it should be scooped out with a hand net. The hook should be removed carefully from the fish. Fish caught can be kept in water-filled containers or keep-nets until the sampling is done.

Hook and line can generate catch per unit effort statistics for particular, targeted groups or species. With mark–recapture methodology it may also be possible to generate more quantitative information on population parameters (Chapter 2).

As a general rule, always exploit local knowledge when using hook and line sampling. This will involve talking to local fishers for advice on rigs and baits. This will also help ascertain if the technique is going to be of use for the species or group of interest. Anglers participating in tournaments may sometimes provide the labour force for a census.

Advantages and disadvantages

Hook and line is cheap and can be a good method to census large, predatory species, species occurring at low density, or those which live in habitats that are difficult to sample by other methods. It is unsuitable for monitoring entire fish communities as a result of its high selectivity (see Biases, below). If a large number of poles and hooks are used, it may quickly become labour-intensive. The chief disadvantage of hook and line fishing is that fish are inevitably damaged and subject to considerable stress. The use of barbless hooks, which can be removed from the mouth more easily, lessens injury to the fish.

Biases

Hook and lining is highly selective, targeting only focal species. The type of bait used will often determine which species is caught. The success of angling (like other passive methods) depends heavily on environmental conditions such as water temperature and on fish behaviour (particularly timing of foraging). It depends also on experience and natural ability.

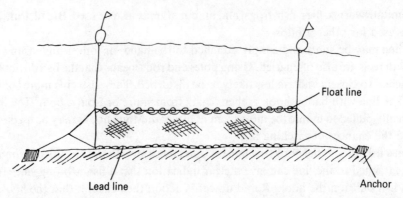

Figure 5.4   Gill net set to sample the lower water column. The relative weights of the float line and lead line can be altered to sample off the bottom at any required depth.

Gill netting

Mobile fish species in freshwater or seawater, under fairly calm conditions

Method

Gill netting is a passive technique which relies on fish trying to swim through diamond-shaped apertures in a net set vertically in the water column. In theory, the apertures are large enough to allow the fish's head through, but not the rest of the body, and the fish becomes trapped as it is unable to back out of the net because of its flaring gill covers. A trammel net is a special type of gill net, which is made of a small-meshed panel hanging loosely between two large-meshed panels. When a fish hits the net from either side, it passes through the large mesh and carries the small-meshed net through the large mesh on the other side, forming a pocket in which the fish is trapped. In practice, with both gill and trammel nets, fish may become wedged across the centre of the body or entangled by fins or spines.

Gill nets usually have a lead line along the bottom edge and a float line along the top (Figure 5.4). By varying the weight of the lead line, gill nets can be made to sink to the bottom, or to stay at the surface of the water, allowing the capture of species at various depths. In shallow water, gill nets can be anchored to the bottom to sample the entire water column.

To deploy a gill net, one end of the net should be attached to a fixed object (buoy, bank, etc.). As the boat moves slowly, the net is fed out by hand or simply allowed to peel out of its storage container. If set too fast, there is little chance of correcting any net entanglement. Once the net is nearly stretched out, the second end is also secured to a fixed point. Do not set the net so tightly across the anchor points that the fish will simply bounce off. In shallow, hard-bottomed waters, gill nets may be set by wading, although further checking and retrieval should ideally be done from a boat. The position of gill nets should be marked with buoys at both ends.

To keep the fish alive, the net needs to be checked regularly while it is set, by leaning over the side of the boat and lifting sections of net to look for gilled fish. When the net has caught enough fish, or after the required amount of time has elapsed, the net is retrieved, starting at the downcurrent or downwind end first, storing each section as it comes in a container or on a gill-net hook. Fish should be removed from the net as they come out of the water. A small hook may be used to lift the mesh over the fish opercula and slide it over the head. Although speed is important to minimise stress to the fish, great care must be taken not to damage fish while freeing them from the net. The net should then be cleared of debris.

For census purposes, gill-net catches are usually expressed as catch per unit effort (CPUE), where effort is calculated as net length × time for which the net is set. Although one may expect catch to increase with net length and set time, beware that these effects are not necessarily linear (Minns & Hurley 1988).

Advantages and disadvantages

Gill netting is a low-cost method of censusing fish. Gill nets are relatively cheap and long lasting, although regular maintenance is necessary. Removing gilled fish and rubbish (detritus, plants, twigs) from the net is time-consuming, particularly if the net is long.

Gill nets are the most selective of nets because mesh size determines exactly the body diameter of fish that will be caught (rather than just the minimum size caught as for other nets). This can be an advantage, if one wants to survey a particular segment of a fish population, or a disadvantage if a more general survey is required. One way to decrease selectivity is to use nets with variable mesh sizes. These nets have a dozen or more mesh sizes arranged in blocks across the net. Another is to use several nets of different mesh sizes. Trammel nets are less selective.

Although there are numerous published studies of selectivity measurements for many species and net sizes, selectivity for the particular net used and the particular species sought may need to be established. This can be very time-consuming. Hamley (1975) gives a thorough discussion of gill-net selectivity.

A major disadvantage of gill netting is that the fish caught in gill nets often die, especially if the net was set for too long or too many fish were caught. If fish are just caught by their gills or perhaps tangled on spines or fins, robust species such as percids may be removed without undue stress and mortality, but sensitive species that lose scales easily (e.g. many cyprinids) are unlikely to survive. Trammel nets tend to damage fish less than gill nets. In addition, gill nets can entangle untargeted animals, including turtles, birds, and mammals. Losing a gill net may be disastrous as it will continue to capture fish and other animals until it finally sinks to the bottom.

Gill nets are most effective in lakes and rivers with little current when the target species is highly mobile. Low visibility increases capture efficiency. In areas of strong water current, gill nets are less effective because the lead line may be lifted off the bottom or the float line dragged down so that the net is not maintained in a vertical position. Wind and waves also make the deployment of a gill net difficult. Gill nets may be set in gaps or channels in vegetated areas, although be prepared for the time-consuming process of removing plant matter from the net when it is lifted!

Biases

Gill nets are highly selective with respect to fish size and also bias the catch in favour of mobile, active, fast-swimming fish. Bottom-dwelling fish are not generally caught. Multifilament gill nets catch slightly larger fish than monofilament or monotwine nets (Hylen & Jakobsen 1979).

Trapping

Most species, under most conditions

Method

Traps for censusing fish fall into three general categories: pot gear, fyke nets (Figure 5.5), and trapping barriers (weirs). However diverse these may be in design and building materials, virtually all traps operate on the 'funnel' or 'maze' principle, with fish passing easily through an entrance hole, but being confused by the blind endings within the trap and being unable to find their way out. Fyke nets exploit this principle further by having a succession of funnels, which concentrates the fish in the final section of the net. They may also have one or two wings of netting (leaders) attached to the first hoop which lead the fish to the entrance of the trap (Figure 5.5b). Pot traps and fyke nets are usually set temporarily, while weirs are more permanent stuctures, often used along sea coasts or rivers to catch migratory species such as salmonids.

Pot traps can be set easily by simply throwing them from boats or from the bank, or by wading in shallow water. Depending on the building material, they may need to be weighted down with rocks or bricks. The position of each trap should be marked with a float, and where the water is deep, the trap should be on a rope to allow it to be easily retrieved. Alternatively, pot traps can be set in midstream and supported on frames to exploit the action of the flow bringing fish to the traps. This is especially useful for migratory species.

Fyke nets require more care in being deployed. When set from a boat, the net should be placed on the bow with the leader(s) on top. The end of the leader is staked or anchored into position, and as the boat moves backward, the leader is let out until it is fully extended. The hoop net is then put overboard, taking care to keep the float line of the leader upright. The net is then stretched out and its end staked in firmly (see Figure 5.5b). Ideally, one of the leaders should extend perpendicular and near to shore to prevent fish from swimming around the trap. Fyke nets may be set singly, in pairs with leaders end to end, or in lines (or 'gangs') with the leader of one net near the end of the previous net.

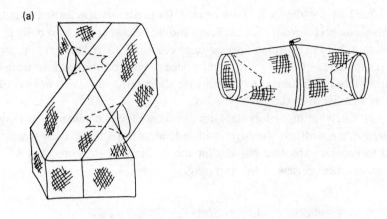

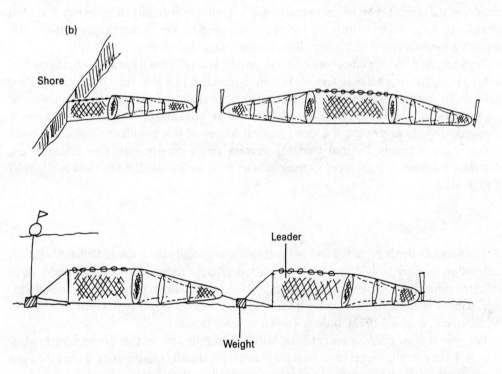

Figure 5.5 (a) Pot traps: artisanal design with two funnels used widely in the Caribbean (left, view from top) and commercially available 'minnow' trap (right), detachable in the middle for easy retrieval of fish. (b) Fyke nets, set singly against the shore (top left), in a pair (top right), or in a gang (bottom), used especially for eels.

Traps are often baited with pieces of fish or meat for predatory species and bread, rice, or fruit for omnivorous/herbivorous species. Traps should be checked at least daily to prevent predation or cannibalism by some species (e.g. eels) and to reduce general stress and mortality. Even for a one-off survey, traps really need to be in position at least from dawn to dusk, and preferably much longer, to account for short-term variations in environmental factors such as weather and water temperature.

With all traps, catch per unit effort statistics can be used with confidence, as long as trap design and size are standardised. Moreover, with individual traps, mark–recapture techniques may be used to generate absolute population size or biomass (Chapter 2, p. 17). Brandt (1984) gives an excellent review of fish trapping.

Advantages and disadvantages

Trapping is one of the most versatile fish censusing methods. It can be used in a wide variety of habitats, from strongly flowing rivers to stillwater lakes and wetlands dominated by vegetation to featureless estuaries, to catch a wide variety of species. It is particularly effective for species occurring at low density or that are active at night. Since traps sample fish passively, the ratio of effort to return can be good. Traps are generally cheap to buy, and even cheaper to make. Because trapping has been used for centuries in most areas of the world, adopting local trapping technology usually saves time and money.

Trapping may become labour-intensive when a large number of traps are used, the catches of fish are high, or regular maintenance of traps is required. In highly vegetated marshy areas, substantial damage to the traps may be caused by mustelids, if present. Traps can also be highly selective (see Biases, below), which can be an advantage, if a particular species is targeted, or a disadvantage, if a more general census of fish populations or community composition is required. Since trapping success varies greatly with trap construction, sampling location, and set time, all these factors must be standardised to yield meaningful CPUE statistics.

Biases

Trap selectivity depends on the size of the trap entrance and size of mesh. Different species, depending on their mobility and activity patterns, territoriality, and inquisitiveness, have different capture efficiencies. Even within a species, trappability is influenced by season, sex, age, habitat availability, etc. Apart from isolated examples (e.g. Perch *Perca fluviatilis* in Windermere, Bagenal 1972), little is known of these biases.

Pot gear is most effective for catching bottom-dwelling species that are seeking food or shelter. Fyke nets also target cover-seeking species, but usually more mobile species than pot traps. Weirs catch mostly migratory species that follow shorelines.

Hydroacoustics

Most fish in deep, fresh and marine waters

Method

Hydroacoustics relies on a sonar system. A transmitter produces a pulse of electricity which is converted by an underwater speaker (the transducer) into an audible sound (usually a 'ping' or 'beep'). This sound travels as a beam through the water and, on hitting an obstacle (the bottom, a fish, or a submarine!), bounces back towards the boat as an echo. This echo is picked up by the transducer and reconverted to electrical signals. This is then modified and amplified by a receiver and displayed on a paper recorder, oscilloscope, or cathode-ray tube.

The strength of the echo (or target strength) is related to the size of the fish, and the length of time taken for the echo to return is related to depth. Hydroacoustic surveys can therefore yield information not only on the number and sizes of fish, but also on their distribution. The numbers of fish present can be estimated by counting the number of returning echoes, if echoes are predominantly from individual fish, or by echo integration, if echoes are from schooling fish. Echo integration is mathematically complex – formulae for its calculations are given by Ehrenberg (1973) – and is greatly affected by errors in estimation of target strengths.

Echosounding equipment is usually installed on the hull of a boat and used vertically to census fish occurring below the boat. It may also be rigged to sample the water column horizontally, which is more useful in rivers, since fish generally orientate against the flow. Surveying can be undertaken from moving boats or from fixed points (a buoy or the bank). The area sampled can be calculated from the dimensions of the beam, although when one is in a moving boat, the duration-in-beam of the target must be calculated from the boat speed. Echosounding allows one to cover the entire area of the water body to be surveyed or to undertake transects (p. 54) or point sampling (p. 55).

Advantages and disadvantages

Although the initial cost of the equipment may be very high, the operational costs of hydroacoustic surveys in terms of ship time and labour are relatively low, giving cost-efficient information on fish density and biomass. Absolute population size can be estimated, even in traditionally difficult locations such as large, fast-flowing lowland rivers, large and deep lakes, and the sea. The technique is non-destructive and fish can be censused *in situ*. Results are obtained quickly, and variance between replicate censuses is generally low.

A serious disadvantage is the inability to distinguish easily between species. Any discrimination relies on detailed knowledge of species distribution, size classes, and composition of stocks as determined by other techniques. The equipment needs to be calibrated in the system to be sampled to correct for environmental noise in the signals.

Bobek (1990) illustrates the differences between different sounding methods, the influence of diurnal activity patterns, and fish orientation on hydroacoustic data.

Hydroacoustic equipment is complex and requires a high level of operator training. To convert reliably echo number into population estimates, target strengths need to be known for all fish likely to be encountered. Evaluating the target strengths of particular species can be difficult and time-consuming (Thorne 1983).

Biases

Fish with swimbladders have target strengths some 10 decibels higher than fish of the same size without swimbladders. This may bias estimated lengths and biomass when species-specific target strengths are unknown. Fish swimming at the surface or sitting on the bottom are not sampled with vertical echosounding. This can be partly alleviated by horizontal sounding. Some fish may also avoid moving boats and swim at a faster speed than can be sampled.

Visual estimates of eggs

Benthic eggs spawned in a mass or a single layer

Methods

Egg number can be assessed visually by a variety of means for species spawning eggs in nests. For species which lay eggs in a single layer (e.g. blennies, gobies, damselfish), the area of the nest covered by eggs can be measured and then converted to egg number if egg diameter is known. For species which lay eggs in a mass (e.g. centrarchids), a scoring system can be devised in which increasing numbers on a scale of, say, 1 to 5 represent increasing numbers of eggs. All nests can then be scored in this way and, if the scale is an absolute rather than a relative one, egg scores may be compared among areas or years. The scores can then be validated by collecting and counting eggs in several (3–5) entire nests of each egg score. Entire nests may be collected with a suction gun (in North America, the devices used for basting turkey work well), large pipettes, or scoops, and the contents transferred underwater into labelled plastic bags. It may also be necessary to collect vegetation, rocks, and debris from the nest as eggs may have adhered to them. The eggs should then be counted the same day in a laboratory.

Advantages and disadvantages

Visual assessment is done quickly and becomes more accurate with experience. Consistency of scoring, both within and among observers, can be a problem. Inter-observer correlations should be examined if more than one individual is collecting the data. Validation of the technique is time-consuming.

Biases

The size of the egg mass relative to the size of the nest may bias visual assessments: small masses in large nests may appear smaller than same-size masses in smaller nests.

Volumetric estimates of eggs

Benthic eggs spawned in a mass

Method

Entire nests may be collected as described under Visual estimates (p. 200), and the volume of the egg mass determined by putting the eggs into a graduated cylinder. Large egg masses may exceed the capacity of the cylinder and may need to be divided. The volume of the egg mass can then be converted into number of eggs by measuring the volume of a small, known number of eggs (50–200, depending on egg size) in the same way.

Advantages and disadvantages

This method is relatively rapid, but in waters deeper than 0.75 m will require the use of SCUBA to collect the eggs.

Biases

The presence of debris to which the eggs can adhere may artificially inflate the estimate.

Plankton nets for catching eggs

Pelagic eggs

Method

Nets used for plankton may also be used for catching floating fish eggs. These nets consist of a funnel of fine-mesh netting held open at the mouth by a circular, metallic ring. The nets are towed, pushed, pulled, or buoyed up, or are held stationary in flowing water (see p. 190). As water is filtered through the mesh, organisms larger than the mesh size concentrate at the end of the net (the 'cod end'). Nets vary tremendously in mouth diameter, but Bowles *et al.* (1978) suggested that nets with a 1-m-diameter mouth sampled pelagic eggs better than any other type of gear.

Monofilament netting with relatively square apertures is generally preferred. The mesh sizes used most commonly to sample fish eggs range between 0.3 and 0.8 mm (up to 1 mm in inland waters). The cod end can consist of a removable bag, jar, or bucket with screened windows. The latter is preferred for collecting fish eggs since it allows more efficient concentrating of the filtrate.

To be able to convert the number of eggs caught into a quantitative measurement, the volume of water filtered must be known. This can be achieved most accurately by attaching a flow meter in the mouth of the net. The rate of flow is then multiplied by the area of the net mouth and the length of time of the tow.

Eggs can be damaged easily; hence extreme care must be taken when transferring the filtrate to a storage container. The eggs can be preserved in a solution of 3–5% formalin (1–2% formaldehyde) in water. Alcohol is not recommended since it causes shrinkage of the eggs, and egg diameter and shape may be important in species identification. The number of eggs caught can then be counted by eye, or with an electronic particle counter.

Advantages and disadvantages

Nets allow a large amount of water to be sampled in a short time. They are also either inexpensive or only moderately expensive. Clogging can be a major problem, leading to passive avoidance of eggs and fish larvae by the net. Performance is also highly dependent on hydrographic characteristics, such as turbulence. Egg density may be used to estimate spawning stock biomass if the sex ratio of the fish population, the proportion of females spawning, and the relationship between fecundity and fish size are known (e.g. Parker 1980).

Biases

Egg densities may appear lower if some eggs were extruded through the mesh under pressure.

Emergence traps for eggs
Benthic eggs

Method

Emergence traps of the type used to catch aquatic insects (see p. 156) may also be used to catch fish larvae hatching from eggs deposited on the bottom. The apparatus is anchored to the bottom over the area of the nest after spawning has taken place. This method is most appropriate for sampling eggs of species, such as salmonids, that scatter or bury their benthic eggs and do not provide parental care lasting until the eggs hatch. Young salmonids may swim down in the gravel and some may be missed unless the edges of the trap are buried deeply.

Advantages and disadvantages

If nests are discrete, each will require its own trap. This has the advantage of providing an estimate of egg number for each nest, but can become expensive since many traps will be required to achieve a good sample size of nests. There is a risk that traps may become dislodged by strong current if left over nests for a long period.

Biases

Traps may not cover the whole area of a nest if the nest is large, thus leading to an underestimate of the number of eggs spawned. This method actually measures the number of eggs hatching rather than the number spawned.

References

Attenborough, D. (1979). *Life on Earth*. Collins/BBC, London.

Bagenal, T. B. (1972). The variability in the number of perch, *Perca fluviatilis* L., caught in traps. *Freshwater Biology* **2**, 27–36.

Bobek, M. (1990). Applied hydroacoustics in cyprinid research. In *Fisheries in the Year 2000*, ed. by K. T. O'Grady, A. J. B. Butterworth, P. B. Spillett & J. L. J. Domaniewski. Institute of Fisheries Management, Nottingham, UK.

Bowles, R. R., Merriner, J. V. & Grant, G. C. (1978). *Factors Associated with Accuracy in Sampling Fish Eggs and Larvae*. US Fish and Wildlife Service, FWS/OBS-78/83, Ann Arbor, Michigan, USA.

Brandt, A. von (1984). *Fish Catching Methods of the World*. Fishing News Books, Farnham, Surrey, UK.

Buckley, B. (1987). *Seine Netting. Advisory Booklet from the Specialist Section-management*. Publications of the Institute of Fisheries Management, Nottingham, UK.

Cowx, I. G. (ed.) (1991) *Catch Effort Sampling Strategies – Their Application in Freshwater Fisheries Management*. Fishing News Books, Blackwell Scientific Publications, Oxford, UK.

Cowx, I. G., Wheatley, G. A. & Hickley, P. (1990). Developments of boom electric fishing equipment for use in large rivers and canals in the United Kingdom. *Aquaculture and Fisheries Management* **19**, 205–212.

Ehrenberg, J. E. (1973). *Estimation of the Intensity of a Filtered Poisson Process and Its Application to Acoustic Assessment of Marine Organisms*. University of Washington Sea Grant Publication WSG 73-2. Seattle, Washington, USA.

Hamley, J. M. (1975). Review of gill net selectivity. *Journal of the Fisheries Research Board of Canada* **32**, 1943–1969.

Hankin, D. G. & Reeves, G. H. (1988). Estimating total fish abundance and total habitat area in small streams based on visual estimation methods. *Canadian Journal of Fisheries and Aquatic Sciences* **45**, 834–844.

Hylen, A. & Jakobsen, T. (1979). A fishing experiment with multifilament, monofilament and monotwine gill nets in Lofoten during the spawning season of Arcto-Norwegian cod in 1974. *Fiskeridir. Skr. Havunders.* **16**, 531–550.

Kennedy, G. J. A. & Stange, C. D. (1981). Efficiency of electric fishing for salmonids in relation to river width. *Fisheries Management* **12**, 55–60.

Minns, C. K. & Hurley, D. A. (1988) Effects of net length and set time on fish catches in gill nets. *North American Journal of Fisheries Management* **8**, 216–223.

Nielsen, L. A. & Johnson, D. L. (eds.) (1983). *Fisheries Techniques*. American Fisheries Society, Bethesda, Maryland, USA.

Parker, K. (1980). A direct method for estimating northern anchovy, *Engravlis mordax*, spawning biomass. *Fisheries Bulletin US* **78**, 5541–5544.

Pereyra, W. T. (1963). Scope ratio-depth relationships for beam trawl, shrimp trawl, and otter trawl. *Commercial Fisheries Review* **25**, 7–10.

Persat, H. & Copp, G. H. (1990). Electricfishing and point abundance sampling for the ichthyology of large rivers. In *Developments in Electric Fishing*, ed. by I. G. Cowx, Fishing News Books, Blackwell Scientific Publications, Oxford, UK.

Smith, P. E. & Richardson, S. L. (1977). *Standard Techniques for Pelagic Fish Eggs and Larva Studies*. Fisheries Technical Paper 175, Food and Agriculture Organization of the United Nations, Rome, Italy.

Thorne, R. E. (1983). Hydroacoustics. In *Fisheries Techniques*, ed. by L. A. Nielsen & D. L. Johnson. American Fisheries Society, Bethesda, Maryland, USA.

Zalewski, M. (1983). The influence of fish community structure on the efficiency of electrofishing. *Fisheries Management* **14**, 177–186.

Zalewski, M. & Cowx, I. G. (1990). Factors affecting the efficiency of electric fishing. In *Fishing with Electricity – Applications in Freshwater Fisheries Management*, ed. by I. G. Cowx & P. Lamarque. Fishing News Books, Blackwell Scientific Publications, Oxford, UK.

6 Amphibians

Timothy R. Halliday

Department of Biology, The Open University, Milton Keynes MK7 6AA,
United Kingdom

Introduction

The habits and life histories of amphibians are such as to pose a number of major problems
for anyone seeking to estimate their numbers accurately. Most are highly secretive in their
habits and may spend the greater part of their lives underground or otherwise inaccessible to
biologists. The limbless caecilians, for example, live entirely beneath the ground surface and
little is known about most aspects of their biology. When amphibians do venture out they
typically do so only at night. They have extremely low food requirements and so can afford to
emerge only when conditions are optimal, typically when the weather is warm and wet. Their
activities are highly seasonal; most temperate amphibians hibernate over winter and many,
notably desert species, aestivate during hot, dry periods.

Amphibians are typically most evident, and thus most easily censused, when they breed,
but breeding activity is characteristically seasonal and may be very unpredictable. In some
temperate amphibians breeding is 'explosive', with annual breeding activity being completed
in one or two days. In such species, effective censusing can be achieved by intensive fieldwork
over a limited period, provided that the censuser is alert to the climatic conditions that
stimulate breeding. In tropical species, however, breeding may occur over an extended
period of the year, sometimes sporadically, so that censusing work has to be maintained over
many weeks or months. In some desert species, breeding does not occur for one or more years
if favourable wet conditions do not occur.

Breeding, in many species, offers good opportunities for censusing amphibians because
they aggregate in large numbers at a limited number of breeding sites, such as ponds. This
applies to those species that lay their eggs in or close to water and have an aquatic larval
stage. There are, however, many amphibians, such as terrestrial-breeding frogs and
salamanders, in which eggs are not laid in water, so that there are no focal points for breeding
activity at which censusing efforts may be directed.

For some species, eggs or larvae are more readily detected, observed, and counted than
adults, and it is tempting to focus censusing efforts on them. This can be highly misleading,
however. In many amphibian species, some individuals, especially females, do not attempt to
breed in a given year. Many amphibians have very high fecundity, but there can be enormous
mortality among eggs and larvae. Amphibians are typically long-lived and breed several
times during the course of their lives and, even in a stable and viable population, there may be
little or no survival of eggs and larvae in a majority of years. Under these circumstances,

Table 6.1. *Summary of the different methods suitable for different groups*
* method usually applicable, + method often applicable, ? method sometimes applicable. The page number for each method is given.

Method	Species that migrate to breed at specific localities	Species that are abundant in a restricted area	Species that are dispersed over a wide area	Page no.
Drift fencing	*			209
Scan searching	*	+	?	212
Netting	*			213
Trapping	*	+		214
Transect and patch sampling		?	*	215
Removal studies	+	*	?	215

counts of eggs and larvae can give a highly misleading measure of a population, especially if they are made only in a single year.

The need to census amphibians has never been more urgent than it is now. Among herpetologists, a growing awareness that amphibians are declining and becoming extinct in many parts of the world, in many instances in areas of apparently pristine and protected habitat, led to the formation in 1991 of the Declining Amphibian Populations Task Force, set up under the auspices of the International Union for Conservation of Nature and National Resources/Species Survival Commission. One aspect of the work of this task force has been the production of a guide to techniques for monitoring amphibian populations (Heyer *et al.* 1994). While that book is largely concerned with methods for assessing amphibian diversity in terms of numbers of species in various habitats, it contains a great deal of information about techniques for censusing and monitoring individual populations.

Recognising individuals

To derive an accurate estimate of a population's size requires, in most censusing techniques, being able to recognise individuals, as in mark–recapture studies (see Chapter 2). The most widely used method for recognising individual amphibians is toe-clipping, but exploiting the fact that most amphibians have highly variable skin markings is becoming more widely used.

Toe-clipping

One or more fingers or toes are removed from an individual, with scissors or toe-nail clippers, using a simple code like that shown in Figure 6.1. It is imperative that, when marking anurans (frogs and toads), the thumb of males is not removed as these bear nuptial pads that are important in mating. Removal of a single digit can be used in mark–recapture studies,

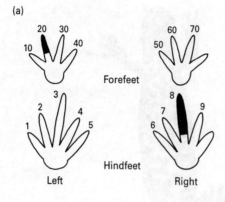

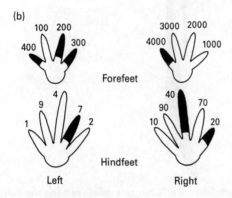

Figure 6.1 Two of a number of schemes used for toe-clipping amphibians. (a) A scheme that limits the number of clipped toes to one per foot, two per animal; this animal is a number 28. (b) A scheme that involves clipping several toes; this animal is number 4967. (*Source*: (a) – author; (b) – Heyer *et al.* (1994) *p*. 280.)

simply to identify individuals that have been caught; removing a different digit in successive seasons yields valuable additional data on aspects of life history such as longevity and frequency of breeding. When multiple-digit codes are used, no more than one digit should be removed from any one limb and no more than two in total from any one individual. Toe-clipping is much more effective as a long-term marking technique for anurans because removed digits do not regenerate; in urodeles (newts and salamanders) digits will re-grow, usually within a year.

The advantages of toe-clipping are that it is simple and inexpensive, and can be used to mark a reasonably large number of animals (up to 99 individuals can be individually marked if not more than a total of two digits are removed). Though widely used, there is increasing concern that it is harmful to amphibians; there is evidence that it decreases survival (Clarke 1972) and that it causes tissue damage (Golay & Durrer 1994). In certain countries it requires

Figure 6.2   Two female Crested Newts (*Triturus cristatus*) showing the belly pattern used in individual recognition. (*Source*: author, from John Baker.)

a licence and anyone doing it should be aware that it causes public disquiet, as in a recent very public scandal in Sweden.

Recognising skin patterns

Several herpetologists have used the natural variation present in the skin patterns of amphibians to recognise up to several hundred individuals. Each individual is photographed or photocopied and entered in a register so that it can be recognised when next caught (Figure 6.2). If a population is large, it is helpful to remove one digit from each registered animal, so that registered and new individuals can be differentiated.

Other methods

A variety of methods have been used to mark and individually identify amphibians; these are summarised in Table 6.2. It should be noted that each method has been used on a limited number of species, in some cases only one, that a method that is effective for one species may not work with another, and that any method needs to be tested and validated for a particular species before it is used on a large scale.

Drift fencing

Amphibians that migrate periodically to small water bodies to mate and/or lay eggs

Method

A fence, supported by posts, is built so as to encircle a breeding site (Figure 6.3). The fence should be at least 35 to 40 cm high and should be dug into the ground to a depth of at least 20 cm. The fence can be made of a variety of materials, including aluminium sheet, plastic sheet, or hardware cloth. Many amphibians are very adept at climbing and it is desirable that the fence has an overhang at the top, at least on the outside, preferably on both sides. The fence should be positioned as near the water as possible, but must be above the anticipated high-water mark. Pitfall taps are dug into the ground, inside and outside the fence, at a spacing of 3 to 5 m along the fence. These may be plastic buckets or metal cans and must be strong enough not to collapse under the weight of the soil around them. To prevent them filling with water, they should have small holes made in the bottom. The pitfalls should also contain cover objects, such as broken flower pots, under which amphibians can hide.

It is vital that pitfall traps are checked frequently: at least once a day, preferably twice. Pitfalls trap predators of amphibians, such as shrews, snakes, beetles, and spiders. In dry weather, amphibians can die from desiccation in pitfalls; in wet weather they may drown in pitfalls that fill with water. Particularly in temperate habitats, most amphibians move towards breeding sites immediately after dark, and an early evening surge of animals is usually followed by a steady trickle throughout the night. The best times to check pitfalls are just after the early evening surge, about 2 to 3 hours after dusk, and as early as possible after daybreak.

It is desirable to back up a drift-fence study with trapping or collecting of animals in the breeding site itself. If all animals caught at the fence are identified in some way, then the ratio of identified to unidentified animals caught inside the fence provides a measure of the effectiveness with which the fence samples the population. However, at some amphibian breeding sites, some individuals may live during the terrestrial phase very close to the water's edge and so may never cross the fence at all.

Advantages and disadvantages

Drift fences can catch a very high proportion of a population, so that, combined with some form of mark–recapture analysis, they can provide a very accurate estimate of population

Table 6.2. *Methods used for marking and recognising individual amphibians*
Note: most of these methods have been used successfully for only one or a few species (see
references) and it should not be assumed that any method is suitable for any given species

Method	Procedure	Advantages	Disadvantages	References
Toe-clipping	One or more digits cut from hands or feet	Inexpensive. Large numbers of individuals can be recognised (not so many if number of digits removed is limited)	Digits of urodeles regrow. May cause tissue damage or reduced survival	Martof (1953)
Elastic waistbands	Fitted to individuals according to their size. Can be colour-coded or numbered	Inexpensive. Large numbers of individuals can be recognised at a distance	Suitable only for short-term studies (a few days or weeks)	Emlen (1968) Davies & Halliday (1979)
Knee-tagging	Small tag tied to knee with stretchable thread	Inexpensive. Large numbers of individuals can be marked	Suitable only for anurans	Elmberg (1989)
Fluorescent colour marks	Coloured fluorescent dust applied to skin with compressed air	Large numbers of individuals can be recognised at a distance and in the dark	Relatively expensive. Marks temporary, lasting 1 to 2 years	Nishikawa & Service (1988)
Skin transplants	Small piece of ventral skin exchanged for piece of dorsal skin	Permanent. Inexpensive	Time-consuming. Requires expertise. Relatively few animals can be marked. Suitable only for species with contrasting dorsal and ventral colours	Rafinski (1977)
Skin staining	Dye sprayed onto skin with dental (panjet) injector	Inexpensive	Can cause injury. Relatively few animals can be marked. Marks short-lived	Wisniewski *et al.* (1980) Gittins *et al.* (1980)

Table 6.2. (*cont.*)

Method	Procedure	Advantages	Disadvantages	References
Tattooing	Dye injected beneath skin with electric tattooer	Inexpensive. Large number of individuals can be marked. Marks last for at least 3 years	Not suitable for dark-skinned species	Joly & Miaud (1989)
Branding	Red-hot or frozen metal applied to skin	Inexpensive	Can cause severe wounds. Marks short-lived in some species	Daugherty (1976) Nace & Manders (1982)
PIT tags	A passive integrated transponder is inserted beneath the animal's skin and a unique radio signal is read with a portable scanner	Very large numbers of individuals can be individually recognised	Very expensive. Suitable only for larger species. Usually requires a licence	Camper & Dixon (1988)
Natural variation in skin patterns	Individuals are photographed or photocopied to provide register of known individuals	Non-invasive	Labour-intensive	Hagstrom (1973)

size. In addition, they yield data on a variety of aspects of the behaviour and life history of amphibians, such as: the direction(s) in which they migrate to and from breeding sites, sex, age and body-size differences in migration times and in time spent at the breeding site, and, if they are kept in place for several years, individual variation in the frequency of breeding. A well-built drift fence will catch all individuals in a population, except of those species that are able to jump or climb over it. The ability of amphibians to find ways across drift fences should not be underestimated and it cannot be assumed that even the best-constructed fence catches every animal.

Drift fences are quite expensive to build; they are labour-intensive, both in building them and in maintaining and checking them. They are easily destroyed by large animals and small children and so are not suitable on farmland or near urban areas unless they are protected by conventional fencing. If not checked frequently, they can cause high mortality among amphibians, through predation, desiccation, or drowning.

More detailed accounts of the design and construction of drift fences, and of the analysis of the data that they yield are provided by Gibbons & Semlitsch (1981) and Heyer *et al.* (1994).

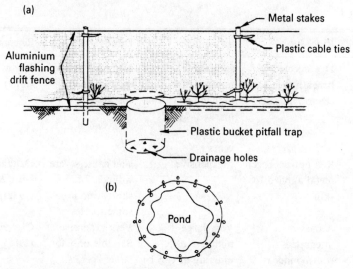

Figure 6.3   A drift fence with pitfall fence around an amphibian breeding site. (a) Position of pitfall traps in relation to the fence. (b) A continuous fence around a breeding pond. (*Source*: Heyer *et al.* (1994) p. 127.)

Biases

Drift fences provide less reliable population estimates of species that are good climbers or jumpers. Very small individuals, e.g. metamorphs, often hide close to a fence and are thus likely to be overlooked. It is possible that, in some species, animals encountering a drift fence will turn away and go elsewhere.

Scan searching

Amphibians that aggregate at breeding sites or that are relatively abundant in a given area

Method

Field workers walk around or through a breeding site or some other prescribed area, systematically searching for animals. The number of animals detected per unit of person-hours provides an approximate estimate of numbers. This is not a suitable method for estimating population size accurately unless combined with a mark–recapture study. For many species, scan searches are best conducted at dusk or at night, using torches, when the animals are most active. For those anurans in which males call, numbers of males may be estimated by counting the number of calling males seen or heard. It should be remembered, however, that, in many species, not all males call in a given time period, and that many frogs stop calling in the presence of humans. In species in which females lay eggs in discrete batches or clumps, female numbers can be estimated by counting egg masses, though it must first be established whether females of the species under study lay one or more clutches in a breeding episode.

Advantages and disadvantages

Scan searching is simple and cheap but is time-consuming.

Biases

Scan searching is subject to two major sources of bias. First, it is a common observation that many amphibians are less wary of humans when they are aggregated in large numbers, so this method can provide an underestimate of numbers when small or dispersed populations are being sampled. Second, males are typically more active than females at breeding sites, may thus be more readily detected, and may thus appear to be more abundant. Furthermore, the two sexes may show very different temporal patterns of arrival at a breeding site, so that repeated sampling over many days or nights is essential.

Netting

Amphibians that aggregate to breed in water bodies

Method

A water body is sampled by means of hand nets or seine nets with the aim of catching as many animals as possible. This is not a suitable method for estimating population size accurately unless combined with a mark–recapture study. Every effort should be made to net all parts of the water body equally thoroughly, but care must be taken not to damage aquatic vegetation which may provide cover and spawn sites for amphibians.

Advantages and disadvantages

Netting is simple and cheap but, for many species, is not very effective as they elude nets quite easily. It can also be very destructive of both the animals and their habitat. It should not be carried out on a large scale when externally gilled larvae are present as they are easily killed by netting.

Biases

Differences in the behaviour and distribution of males and females while at a breeding site may lead to very different numbers of the two sexes being caught, leading to a spurious estimate of the sex ratio.

Trapping

Amphibians that aggregate at breeding sites or that are relatively abundant in a given area

Method

Traps are set out in a water body or over an area of ground and are checked frequently. Since amphibians eat only live food, it is generally impractical to bait traps; trapping relies instead on the fact that many amphibians move extensively around their habitat when the weather is suitable. Trapping is not a suitable method for estimating population size accurately unless combined with a mark–recapture study. In water bodies, a variety of traps suitable for fish are widely used (see Chapter 5). Since many species need to go to the water surface to breathe, it is essential that traps are either only partially submerged or that they provide some kind of access to the surface, especially in warm weather. Small amphibians and larvae can be sampled by simple traps made out of plastic drink bottles (Griffiths 1985). Terrestrial amphibians can be sampled by pitfall traps similar to those used at drift fences (see above) and by setting out a number of artificial cover objects such as pieces of wood or paving slabs.

The positioning of traps can be crucial and they are much more successful if they are positioned so as to intercept amphibians during regular daily movements. For example, many aquatic salamanders and newts move from deep water during the day to shallower pond margins at night. Funnel traps are thus most effective if they are set so that their entrance points towards deep water at dusk, and towards the bank at dawn.

Advantages and disadvantages

Trapping is simple and cheap but can be labour-intensive. Traps typically provide a low return, in terms of the number of animals caught, unless a population is particularly dense. Mortality can occur in traps, by lack of oxygen, overheating, or predation, and it is essential that traps are checked regularly and that animals do not remain long in traps on warm days.

Biases

Differences in the behaviour and distibution of males and females while at a breeding site may lead to very different numbers of the two sexes being caught, leading to a spurious estimate of the sex ratio.

Transect and patch sampling

Terrestrial amphibians

Method

Transect sampling is most effective if combined with other methods described above, such as drift fencing or trapping. A line is marked out over an area of ground, and fieldworkers walk along it at regular intervals, systematically searching for animals. Cover objects within a specified distance of the line (e.g. 2 m) are turned over and checked, as are any traps set along the transect. The number of animals detected per unit of person-hours provides an approximate estimate of numbers. This is not a suitable method for estimating population size accurately unless combined with a mark–recapture study. A particular strength of transect sampling is that it can be used to relate amphibian abundance to habitat variables, such as altitude and vegetation.

Patch sampling is essentially the same method, but involves systematically traversing and searching one or more marked-out patches rather than a linear transect. The patches may take the form of quadrats, a series of small squares laid out at randomly selected sites within a habitat. When placed randomly, and in sufficient number, quadrat samples enable statistical inferences to be made about the abundance of amphibians and their distribution in relation to habitat variables.

Advantages and disadvantages

These are cheap and simple methods, but are labour-intensive. They may yield a low return in terms of numbers of animals found, except where a species is particularly abundant. Quadrat sampling is especially suitable for species living on the forest floor (Heyer *et al.* 1994).

Biases

Transects or sample patches may not cross habitats of species with very specific habitat requirements, but researchers should avoid selecting areas that 'look good' for amphibians. Individuals of very active species may escape before they can be sampled, and their numbers may therefore be underestimated.

Removal studies

Amphibians that aggregate at breeding sites or that are relatively abundant in a given area

Method

A water body or a marked-out area of ground is systematically searched, netted, or trapped and all animals found are removed and held elsewhere, to be returned subsequently. This process is repeated at predetermined intervals. The ratio of numbers caught in successive captures is used to estimate the population size (Heyer *et al.* 1994).

Advantages and disadvantages

This is a cheap and simple method, but is labour-intensive. If a pond or area of ground is to be sampled thoroughly, it can be very destructive of the animals' habitat, and ponds should not be subjected to such treatment when eggs or larvae are present. Animals removed from a site must be kept under suitable conditions until they are returned. Even if no mortality occurs, the successful reproduction of animals temporarily removed from the field can be compromised; some amphibians show a marked stress reaction to capture and quickly go out of breeding condition. An advantage of removal studies is that, while animals are held in captivity, much useful data can be collected, such as age, sex, and breeding condition.

Biases

This method assumes that animals are equally catchable at all times; because the behaviour of amphibians varies considerably with changes in the weather, this assumption is rarely true.

References

Camper, J. D. & Dixon, J. R. (1988). Evaluation of a microchip marking system for amphibians and reptiles. Research Publication 7100–159, Texas Parks & Wildlife Dept., Austin, Texas.

Clarke, R. D. (1972). The effect of toe-clipping on survival in Fowler's Toad (*Bufo woodhousei fowleri*). *Copeia* 182–185.

Daugherty, C. H. (1976). Freeze-branding as a technique for marking anurans. *Copeia* 836–838.

Davies, N. B. & Halliday, T. R. (1979). Competitive mate searching in the Common Toad *Bufo bufo*. *Animal Behaviour* **27**, 1263–1267.

Elmberg, J. (1989). Knee-tagging – a new marking technique for anurans. *Amphibia-Reptilia* **10**, 101–104.

Emlen, S. T. (1968). A technique for marking anuran amphibians for behavioral studies. *Herpetologica* **24**, 172–173.

Gibbons, J. W. & Semlitsch, R. D. (1981). Terrestrial drift fences and pitfall traps: an effective technique for quantitative sampling of animal populations. *Brimleyana* **7**, 1–16.

Gittins, S. P., Parker, A. G. & Slater, F. M. (1980). Population characteristics of the Common Toad (*Bufo bufo*) visiting a breeding site in mid-Wales. *Journal of Animal Ecology* **49**, 161–173.

Golay, N. & Durrer, H. (1994). Inflammation due to toe-clipping in Natterjack Toads (*Bufo calamita*). *Amphibia-Reptilia* **15**, 81–96.

Griffiths, R. A. (1985). A simple funnel trap for studying newt populations and an evaluation of trap behaviour in Smooth and Palmate Newts, *Triturus vulgaris* and *T. helveticus*. *Herpetological Journal* **1**, 5–10.

Hagstrom, T. (1973). Identification of newt specimens (Urodela *Triturus*) by recording the belly pattern and a description of photographic equipment for such registration. *British Journal of Herpetology* **4**, 321–326.

Heyer, W. R., Donnelly, M. A., McDiarmid, R. W., Hayek, L.-A. C. & Foster, M. S. (eds.) (1994). *Measuring and Monitoring Biological Diversity. Standard Methods for Amphibians.* Smithsonian Institution Press, Washington DC.

Joly, P. & Miaud, C. (1989). Tattooing as an individual marking technique in urodeles. *Alytes* **8**, 11–16.

Martof, B. S. (1953). Territoriality in the Green Frog, *Rana clamitans. Ecology* **34**, 165–174.

Nace, G. W. & Manders, E. K. (1982). Marking individual amphibians. *Journal of Herpetology* **16**, 309–311.

Nishikawa, K. C. & Service, P. M. (1988). A fluorescent marking technique for individual recognition of terrestrial salamanders. *Journal of Herpetology* **22**, 351–353.

Rafinski, J. N. (1977). Autotransplantation as a method of permanent marking for urodele amphibians (Amphibia, Urodela). *Journal of Herpetology* **11**, 241–242.

Wisniewski, P. J., Paull, L. M., Merry, D. G. & Slater, F. M. (1980). Studies on the breeding migration and intramigatory movements of the Common Toad (*Bufo bufo*) using Panjet dye-marking techniques. *British Journal of Herpetology* **6**, 71–74.

7 Reptiles

Simon Blomberg

Department of Anatomical Sciences, University of Queensland, St Lucia, Queensland 4072, Australia

Richard Shine

Zoology A08, School of Biological Sciences, University of Sydney, Sydney NSW 2006, Australia

Table 7.1. *Summary of the different methods suitable for different groups*

* method usually applicable, + method often applicable, ? method sometimes applicable. The page number for each method is given.

Method	Snakes	Lizards	Crocodilians	Turtles and tortoises	Page no.
Hand capturing	*	*	*	*	219
Noosing		+			221
Trapping	?	+	?	+	222
Marking individuals	*	*	*	*	224

Introduction

Most common survey methods employed to estimate the abundance of reptiles involve capturing individuals. This is for two reasons: (a) reptiles tend to be mobile and/or shy and cryptic, so that not all members of a population will be visible (and therefore amenable to counting by sight) at any one time; and (b) much more information can be obtained from an animal that has been captured than can be obtained from an animal that has simply been seen. For example, the animal may be weighed and measured, have its sex and reproductive condition determined, and have its parasite load assessed. An identifying mark may also be placed on the animal so that it can be re-identified, should it be recaptured at a later time. As well as providing advice on capturing reptiles, we also provide information on common techniques for marking individuals.

Reptiles are ectotherms. That is, they obtain their body heat from the external environment. This has major implications for any survey technique, in that weather conditions may greatly affect the activity, and therefore the catchability of reptiles. The effect of weather can vary seasonally, as well as on a daily basis. This should be kept in mind when designing a survey programme.

Because it is unlikely that the whole population will be counted in any one census period (some individuals will be missed), statistical mark–recapture methods should generally be used to estimate population size/densities or survival probabilites for reptile populations (see Chapter 2).

Hand-capturing

Small terrestrial snakes, lizards, and tortoises

Method

The simplest method used by field herpetologists to capture lizards (and other small terrestrial reptiles) is to search intensively in microhabitats which they are known to frequent, and catch them by hand. Small lizards and snakes are found most easily by looking in potential shelter sites, for example by turning over rocks and logs, or by stripping bark from trees (for arboreal species). Sheets of scrap metal and derelict car bodies are often fruitful places to look (because the sun-warmed metal is very attractive to many reptiles), though they could not be described as 'natural' habitats! Some herpetologists have standardised their 'catching effort' by placing their own consistently sized, regularly spaced sheets of tin in appropriate sunny positions, and looking under them on some consistent schedule.

When 'turning cover', a pair of gardening gloves can be worn to prevent cuts to the hands, as well as bites and stings from various arthropods. If there are venomous snakes in the survey area, care should obviously be taken to avoid getting bitten. Logs and rocks should be turned towards the fieldworker, so that the log or rock is between the fieldworker and the snake.

A small hand-held torch powered by one or two penlight batteries is invaluable for looking into cracks or holes in search of reptiles. Alternatively, a mirror with a small hole in the centre to look through can be used. This has the advantage that shadows cast by the observer will not obscure the view.

When turning rocks or logs, a hand-held rake is useful for searching for small lizards and snakes that may have crawled under the leaf litter; however, one can improvise by using a small stick. Geckos often cling upside-down to rocks, so both the undersurface of rocks and the leaf litter should be examined where geckos occur. All logs and rocks should be replaced in their original position, so that disturbance to the habitat is minimised.

Some of these techniques may be particularly destructive to the habitat, both for the reptiles and for other organisms, so care should be taken to avoid undue disturbance. Care should also be taken when designing the survey to ensure that the survey method will not affect the habitat in such a way that subsequent survey estimates may be biased. Some species may increase in abundance in response to habitat disturbance, and some may decrease in abundance. This is especially important when censusing rare or endangered reptiles. One would certainly not want the survey method to cause the extinction of the study species, or any other species! Be sure to obtain any necessary permits from the appropriate wildlife and land management authorities before you begin your survey.

Many lizards and snakes can also be counted or captured during their activity periods, rather than when they are hiding under cover. For many diurnal species, mid-morning is a good time to search, while the reptile is basking to elevate its body temperature. Walk slowly with the sun at your back, pausing frequently to scan suitable microhabitats for your quarry. Nocturnal species can often be found by torchlight, in the same way as is often used for amphibians (see Chapter 6). Some species, such as geckos, are best detected by their dull eye shine (the reflection of the light source in the animal's eyes). This is easiest with a dull white light. Some geckos (e.g. *Diplodactylus maini*) fluoresce under ultraviolet light, and can be located using a small hand-held UV lamp (C. Dickman, personal communication).

Having located a lizard or a non-venomous snake, the easiest way to catch it is to simply pounce on it with an open, cupped hand, taking care not to crush it. Care should also be taken with species that practice tail autotomy (drop their tails). These species should not be held by the tail. Gloves can be useful when catching lizards or non-venomous snakes using this method, as even these animals can often give a nasty bite.

If a venomous snake has been located, it can be caught by pinning it behind the head, using a Y-shaped stick with some padding in the fork. The snake can then be picked up, with the neck held firmly (but not so tight as to choke the snake) between thumb and forefinger. Some larger species may also be picked up by the tail, and then pinned down by the neck. 'Snake tongs', large forceps manipulated by a trigger-grip, may also be useful when catching snakes. Sticks without padding should not be used to pin snakes, as they may hurt the snake. The handling of venomous snakes is a skill that has to be learnt. Fieldworkers wanting to catch snakes should practise handling non-venomous snakes first, and should have a good knowledge of the necessary first-aid procedures. First-aid procedures differ for the different families of venomous snakes, so fieldworkers should be aware of the appropriate procedures for the bites of any snakes they are likely to encounter. No fieldworker should conduct such ecological surveys in remote areas on their own if dangerous snakes are known to be present.

Once one has caught the lizard or snake, data can be taken from it immediately and it can be released, or it can be placed in a bag with a label for 'processing' at a later time. Cloth bags can be used for large specimens, and plastic bags can be used for small ones. If using plastic bags, make sure the bag is inflated with air, and contains some leaf litter for cover. More than one individual can be kept in a single bag; however, aggressive or cannibalistic species should be kept separately. Reptiles should not be kept in bags for more than a few hours, and they should never be left in the sun or inside a parked vehicle, because they will quickly succumb to heat exhaustion.

Terrestial tortoises can be found and captured in the same way as lizards, but aquatic species (terrapins and turtles) obviously require different techniques. Freshwater species living in clear water can often be caught while snorkelling, although some soft-shelled turtles are too fast moving for even the strongest human swimmers. One alternative hand-capture technique for freshwater turtles is the use of a long-handled dip net from a canoe. It takes lots of practice, but it does work. Marine turtles are usually censused when females come ashore to lay eggs. Hatchling sea turtles may be censused when they leave the nest and migrate across the beach to the ocean. Male and immature sea turtles can be captured by hand by diving off a small power boat; however, this method takes some practice.

Crocodilians (crocodiles and alligators) are usually hand-caught at night, because their eyes are highly reflective to the torch beam, allowing them to be located easily. Smaller crocodiles (< 1.5 m) are simply jumped upon (beware of their larger relatives as you do so!), while larger animals are usually caught with small harpoons that lodge in the animal's scales. Some scientists catch alligators with heavy-duty fishing rods, simply casting a large hooked lure behind the alligator and retrieving it rapidly, so that the hooks lodge in the alligator's armour. Even small crocodilians can cause considerable lacerations if they are able to bite, so great caution and prior training is essential.

Advantages and disadvantages

Hand-capturing small lizards and snakes requires little specialised equipment, and many common species can be located and captured relatively quickly and easily. Retreat sites can be located and characterised. Hand-capturing is, however, labour-intensive. Suitable cover must be present if reptiles are to be captured from shelter sites. It is also difficult to standardise capture effort between fieldworkers. This method can also disturb the habitat considerably, if care is not taken to replace cover in its original position.

Biases

For terrestrial species, this method is biased towards species that use logs, rocks, bark, etc. for cover, and is not a good method for catching species that do not use these microhabitats, for example burrowing species. Animals that live under rocks or logs that are too heavy to lift are also not sampled.

Some species may show sex- or size-specific differences in catchability, which may be caused by sex- or size-specific differences in habitat use or activity patterns. However, it may be possible to alter the survey design to take such differences into account.

Noosing

Many lizards

Method

Many lizards are most visible when they are active. However, they are wary when approached, and evade hand-capture by running away. Also, some lizards, such as varanids, iguanids, and agamids, may sleep or bask in places that are difficult to reach, such as in the canopy of tall trees, or in burrows. In these situations, a noose may be used to capture lizards without having to approach the lizard too closely. Nooses consist of a long pole, with a loop of string at the tip, which can be tightened around the neck of a lizard and pulled tight in order to capture the animal. Nooses for small species may be constructed at short notice, using a small length of fishing line or dental floss, and a stick. Heavy-duty nooses can be constructed from a long, strong fishing rod and some lightweight cord. These are especially useful for large agamids, iguanids, and varanids.

A sophisticated rubber-band-powered noosing 'gun' may be constructed from a fishing rod 'blank', with a noose made of fishing line attached to the tip. Fishing line is threaded through the hollow centre of the fishing rod, and attached to a trigger grip. When the trigger is pulled, a rubber band tightens the noose (see Bertram and Cogger 1971 for construction details). Care must be taken so that there is not too much tension on the noose, or the lizard may be injured.

To use a noose, the lizard must be approached slowly, and the noose slipped over the lizard's head. The lizard will frequently not be disturbed by this. In fact, many species will attempt to bite the end of the noose! When the noose is around the neck of the lizard, the string or fishing line should be quickly pulled tight. Lizards will frequently struggle violently, so the noose should be loosened and the lizard removed as soon as possible.

A related technique that works well for tropical skinks (for example, *Emoia*), and which is the favoured technique for catching small lizards in the Pacific islands, involves attaching a small insect to a 'fishing rod' made from a stick and some fishing line. The insect is dangled in front of the lizard, which will invariably attack and try to eat it. Once the lizard has a firm grasp on the insect, the lizard is flicked up into the air using the rod, and is caught with the free hand.

Advantages and disadvantages

This technique is good for catching active lizards, or lizards in places that are difficult to reach, such as at the entrance to burrows, or in trees. It is useful for species that are difficult to approach when active. Noosing works best with lizard species with distinct necks. Simple nooses can be easily constucted when needed. It is, however, labour-intensive and time-consuming, and thus faces similar problems to simple hand-capturing.

Biases

Noosing is biased towards active lizards that can easily be located by sight. It is not a useful technique for cryptic or burrowing species.

Trapping

Many small reptiles

Method

For terrestial reptiles, the most commonly used trap is a pitfall trap, consisting of a bucket sunk into the ground so that the lip is flush with the surface. A small layer of leaf litter or some other cover should be provided for animals that fall in. This may also have the benefit of attracting arthropods to the trap, which can act as bait for lizards.

Cans, plastic tubes, or even milkshake cups can be used in place of buckets; the diameter and depth of the trap may affect the size and species composition of the reptiles that are trapped, so some experimentation with pitfall trap design may be necessary. For example, James (1989, 1991) used buckets (330 mm in diameter × 450 mm deep) in a study of seven sympatric skink species. Adult body sizes ranged from 35 mm snout–vent length (SVL) to 100 mm SVL. Traps may be left in place permanently, or removed between census periods. If the traps are left in place, lids should be fitted to the traps so that no animals can fall in between census periods. If pitfall traps are used in a forest or woodland, the traps can be filled with sticks so that any animal that enters the trap between censuses can crawl out by climbing up the sticks. In high-rainfall areas, pitfall traps should have drainage holes in the bottom, so they do not fill with water. The drainage holes should not be so large, however, that animals can crawl through them and live under the trap!

Pitfall traps can be placed next to logs, rocks, or other habitat features that may be frequented by lizards or snakes. Alternatively, pitfall traps may be used in conjunction with drift fences to increase capture success. Drift fences are low fences (approximately 30 cm high) made of polythene or some other cheap flexible material, and held erect by pegs. The bottom of the fence is dug a few centimetres into the soil to prevent animals burrowing underneath. Drift fences run along a trap line, between the traps, acting to guide reptiles into the traps. Reptiles that are crawling through the survey area come up against the fence and follow it along, looking for a way around the fence. They eventually come to a pit trap, and fall into it. It is important to use a material for the drift fence that is difficult for lizards and snakes to climb. Nothing is so depressing as walking along a drift fence and seeing all the lizards that you wished to catch sitting on top of the fence, basking in the sun, and then jumping off and running away as you approach!

Traps should be checked and animals processed at least once per day during a trapping study. Traps may have to be checked more frequently in harsh environments (e.g. deserts) to avoid stress and mortality of trapped animals, or where trap success is particularly high. The spacing and size of pit traps and the use of drift fences for designing sampling programmes has been discussed by Friend *et al.* (1989) and Morton *et al.* (1988). Check your traps carefully with a stick or gloved hand before reaching in to pick up that lizard – there may also be a hidden snake, spider, or scorpion!

An alternative to pitfall traps that can be used in rocky areas where holes cannot be dug is the wire funnel tap (Fitch 1987). Made in the same shape as a lobster trap or fish trap, it is basically a cylinder with a funnel at one or both ends. A drift fence leads the reptile into the funnel, and thence into the trap. A wide range of sizes is possible, depending on the species to be caught. Large species require strong wire-mesh traps, whereas smaller animals can be caught in plastic bottles with the funnel attached to the lid.

Baited funnel traps (Legler 1960) or drum nets are very effective for catching many species of freshwater turtle. The bait is usually meat or fish (cans of sardines are quick and easy), and the trap must be set so that part of it projects above the water. This gives trapped turtles a chance to breathe – otherwise, they will drown.

One point to remember with any trap you set is that you have the responsibility to check the trap regularly. A lost or abandoned trap could go on catching (and killing) animals for many decades.

Advantages and disadvantages

Trapping often allows the capture of a large number of individuals, with comparatively little effort from the investigator. Trapping effort can be standardised in space and time, satisfying the assumptions of many statistical models for estimating survival or population size. Traps and drift fences may have to be constructed and installed at the survey site, which may take considerable time, effort, and money. Traps have to be checked regularly, and closed or removed when not in use. Lost or abandoned traps may cause the death of animals at the survey site long after the study has concluded.

Biases

Only animals that are actively moving within the survey area will come into contact with traps or drift fences. Therefore, trap success will generally be very low in poor weather, when reptiles are inactive. Highly sedentary species will not be sampled.

Marking individuals

All reptiles

Having captured a reptile, and if a mark–recapture procedure is to be used to estimate the abundance of animals or their probability of survival over some time period, it is necessary to give individuals a unique mark so that they may be identified upon recapture. As with capture methods, the particular technique you adopt will depend on the natural history of the study species, so some experimentation will probably be necessary, and usually desirable.

Method

Marks may be applied to reptiles so that they last for short time periods, for longer time periods, or permanently. Marks can be applied using some sort of coding system, so a large number of combinations of marks can be used (Woodbury 1956).

Perhaps the simplest technique is to paint a mark on each animal, using paint (nail polish works well). A colour-coding system can easily be devised.

The most common technique that is used to mark small lizards is to remove toes from one or more feet using sharp fingernail scissors. Feet and toes can be numbered, and can yield a large number of combinations, depending on the coding system. Permanent marks can be applied to lizards or snakes by clipping scales and cauterising them with a small hand-held soldering iron. The scales will either not grow back, or will grow back in a different colour (usually white) if the underlying pigment layer has been destroyed. An alternative method is to mark animals using a hot or cold branding iron. Turtles and tortoises can be marked by cutting notches in the edge of the carapace. Crocodiles and alligators can be marked by removing tail scutes. Reptiles may also be tattooed using a small tattoo gun.

Passive integrated transponder (PIT) tags have recently become available for marking animals, and promise to be an effective method for permanently marking reptiles (Camper and Dixon 1988, Keck 1994). Tags can be inserted under the skin or in the abdominal cavity of small specimens. Each tag has a unique code that can be read using a hand-held scanner.

Advantages and disadvantages

Paint marking has the advantage that the animal is not physically harmed in any way. However, it may make animals more visible to predators. Also, paint will wear off after several days, so this technique is only useful for very short-term studies.

The natural history of the species should be taken into account when deciding on an appropriate marking procedure. Toe-clipping would be a cruel and inappropriate procedure to use for highly arboreal lizards, or those with thick, fleshy toes. However, many terrestrial species seem to be unaffected by toe-clipping.

Apart from paint marking, the methods mentioned above all involve physically altering animals in some way. These methods result in a mark that is longer lasting or permanent; however, care must be taken so that animals are not harmed by the marking procedure. Permanent marking techniques, such as tattooing or toe-clipping, should only be carried out by experienced, trained herpetologists or under the supervision of a veterinarian. Fieldworkers should be aware of ethical considerations when censusing reptile populations, as accurate censusing requires interfering with animal populations to a certain extent. Study methods should be submitted to the appropriate animal ethics committee for approval, when required by law.

Biases

Marks that are ambiguous or are lost can make the analysis of mark–recapture data very difficult. A major assumption of mark–recapture models is that marks are unique, and most models assume that marks are not lost over time. Another common assumption is that marking individuals does not affect their behaviour. If these assumptions are not met, population size estimates and survival probabilities will be severely biased. Loss of marks can be corrected for by marking individuals twice, and then calculating the rate of loss. A better solution is to develop a better marking technique for your species of interest.

Acknowledgements

We thank M. Olsson and C. Dickman for useful comments on previous drafts of this chapter.

226   *Reptiles*

References

Bertram, B. P. & Cogger, H. G. (1971). A noosing gun for live captures of small lizards. *Copeia* 1971, 371–373.

Camper, J. D. & Dixon, J. R. (1988). Evaluation of a microchip marking system for amphibians and reptiles. Research Publication 7100–159, Texas Parks and Wildlife Department, 1–22.

Fitch, H. S. (1987). Collecting and life-history techniques. In *Snakes Ecology and Evolutionary Biology*, ed. by R. A. Seigel, J. T. Collins & S. S. Novak, pp. 143–164. Macmillan, New York.

Friend, G. R., Smith, G. T., Mitchell, D. S. & Dickman, C. R. (1989). Influence of pitfall and drift fence design on capture rates of small vertebates in semi-arid habitats of Western Australia. *Australian Wildlife Research* 16, 1–10.

James, C. D. (1989). Comparative ecology of sympatric scincid lizards (*Ctenotus*) in spinifex grasslands of central Australia. Unpublished Ph.D. Thesis. University of Sydney.

James, C. D. (1991). Population dynamics, demography and life history of sympatric scincid lizards (*Ctenotus*) in central Australia. *Herpetologica* 47, 194–210.

Keck, M. B. (1994). Test for detrimental effects of PIT tags in neonatal snakes. *Copeia* 1994, 226–268.

Legler, J. M. (1960). A simple and inexpensive device for trapping freshwater turtles. *Proceedings of the Utah Academy of Science* 37, 63–66.

Morton, S. R., Gillam, M. W., Jones, K. R. & Fleming, M. R. (1988). Relative efficiency of different pit-trap systems for sampling reptiles in spinifex grasslands. *Australian Wildlife Research* 15, 571–577.

Woodbury, A. M. (1956). Uses of marking animals in ecological studies: marking amphibians and reptiles. *Ecology* 37, 670–674.

8   **Birds**

David W. Gibbons

The Royal Society for the Protection of Birds, The Lodge, Sandy, Bedfordshire
SG19 2DL, United Kingdom

David Hill

Ecoscope Applied Ecologists, 9 Bennell Court, Comberton, Cambridge CB3 7DS,
United Kingdom

William J. Sutherland

School of Biological Sciences, University of East Anglia, Norwich NR4 7TJ,
United Kingdom

Introduction

Birds are perhaps the easiest of animals to census. They are often brightly coloured, relatively easy to see, and highly vocal. They are also very popular to study, with the result that there are high-quality field guides available and many professionals and amateurs with a high level of identification skills. Because of this popularity, they are undoubtedly the most frequently censused of all taxa. Though some may argue that birds receive more than their fair share of the monitoring cake, the widespread involvement of volunteers in many schemes makes bird census and monitoring an extremely cost-effective way of monitoring the overall health of the environment (Furness & Greenwood 1993).

Birds often give away their presence vocally, and many species of birds are best detected by their calls and songs (e.g. Rappole *et al.* 1993). There are, however, some potential pitfalls in using song as a census tool. These problems affect several of the methods outlined below, particularly territory mapping, point counts, and line transects. Only territorial males usually sing and the non-breeding part of the population, which may be substantial, is very difficult to census. Birds may also call less at low densities (La Perriere & Haugen 1972). Unpaired males of some species sing more than paired males. For example, unpaired male Ovenbirds *Seiurus aurocapillus* sing 3.5 times more often than paired males, while unpaired Kentucky Warblers *Oporonis formosus* sing 5.4 times more often than paired males (Gibbs & Wenny 1993). Male Sedge Warblers *Acrocephalus schoenobaenus* cease to sing as soon as they have attracted a mate, while males of the cogeneric Reed Warbler *Acrocephalus scirpaceus* continue to sing once mated (Catchpole 1973). Such differences between species thus necessitate a knowledge of the natural history of the species in question, else population estimates may be distorted and investigations of suitable breeding habitat confused.

Bibby *et al.* (1992) give a good review of techniques and show how they can be applied to (mainly) European birds, while Koskimies & Väisänen (1991) describe in detail the

Table 8.1. *The use of different methods for different groups of birds.*
∗ method usually applicable, +method often applicable, ? method sometimes applicable. The page number for each method is given.

Method	Water-birds	Sea-birds	Wading birds	Birds of prey	Game-birds	Near passerines	Passerines	Page no.
Counting nests in colonies	+	∗				+	?	229
Counting leks					∗		?	232
Counting roosts	+		∗				?	234
Counting flocks	+		+			?	?	235
Counting migrants				+			?	236
Territory mapping	+		+	+	+	+	∗	238
Point counts	?		?	?	?	+	∗	243
Line transects	+	∗	+	+	+	+	∗	245
Response to playback		+		+		?	?	249
Catch per unit effort (mist netting)						?	+	250
Mark–release–recapture	?	?	?	?	?	?	?	252
Dropping counts	+				+			253
Timed species counts						?	+	254
Vocal individuality	?	?	?	?	?	?	?	255

methods used in Finland. Verner (1985) and Dawson (1985) critically review many census methods, while the proceedings of the Asilomar Symposium (Ralph & Scott 1981) and of the International (now European) Bird Census Council conferences (e.g. Purroy 1983, Taylor *et al.* 1985, Blondel and Frochot 1987, Haila *et al.* 1987, Stastny & Bejcek 1990, Hagemeijer & Verstrael 1994) contain many individual research papers on census techniques. More recent reviews are presented by Baillie (1991) and Spellerberg (1991), while Terborgh (1989) provides a very readable critique of the major North American bird monitoring schemes. Buckland *et al.* (1993) is the most authoritative text on point counts and line transects.

Table 8.1 lists the methods that have been used for censusing birds. Though several methods are listed, there are broadly two types: those for censusing species that are evenly distributed across the landscape, and those for species that are not (i.e. are highly clumped). For example, territory mapping, point counts, and line transects are best for species which are evenly distributed (e.g. territorial species), while counts of colonies, roosts, flocks, and leks are best for species with clumped distributions. In addition, the dispersion of a species may vary throughout the annual cycle. For example, males of many species of grouse gather at leks early in the breeding season, but are widely dispersed at other times; seabirds often gather at breeding colonies in the spring, but are out at sea for much of the rest of the year. It is often much simpler to count birds when clumped than when dispersed.

Counting nests in colonies

Colonial nesters, particularly seabirds and herons, some passerines and near-passerines

Method

About one-eighth of bird species nest in colonies and as a consequence are particularly easy to census during the breeding season. The technique adopted depends upon whether the colony is on a cliff face, or whether the species nests on the ground, in burrows, trees, or bushes. Each is treated in turn below, and each relies upon discriminating occupied from unoccupied nests. Birkhead & Nettleship (1980) and Lloyd *et al.* (1991) provide details of methods for counting colonial seabirds. Walsh *et al.* (1995) provide an extremely comprehensive suite of methods for censusing British and Irish seabirds.

Many colonial nesting species breed synchronously. This is advantageous for censusing as it means that all breeding birds will be at the colony during the same period and at a similar stage in the nesting cycle. The best time to count is generally from midway through incubation to early in the nestling stage (Bullock & Gomersall 1981, Hatch & Hatch 1989). Any earlier and some clutches may not have been started, any later and some pairs may already have lost chicks and deserted the nest site, both leading to underestimates of the total breeding population. For colonial species with a more protracted breeding season and with high rates of nest failure, e.g. Greater Flamingo *Phoenicopterus ruber*, population estimation can be more difficult (Green & Hirons 1988).

Cliffs

Ideally count from a position slightly above but opposite the colony. Many ornithologists have died studying seabirds and it is important to ensure that the counting position and access route to it are safe; loss of life is more serious than loss of data. Ideally pairs of birds or occupied nest-sites (or at least apparently occupied nest-sites) should be counted. For some species which nest at very high densities (e.g. Guillemot *Uria aalge*), however, pairs of birds are difficult to count as this requires identification of all sites with eggs, young, or incubating adults; this can take hours of observation (Lloyd *et al.* 1991). For such species, counts of individual birds are more effective. To aid in counting, it is advisable to divide the colony up into a number of subunits. Colony attendance can vary both with season and diurnally and should be taken into account in deciding when to count.

For some highly visible colonial nesters, particularly those which, like the Gannet *Morus bassanus*, build substantial nests, it may be simpler to photograph the colony and count nests directly from the photograph. This method is particularly useful where there is no suitable position to count from, yet a photograph could be taken (e.g. from a boat or from the air).

For some species, photography may be unsuitable, because black and white seabirds are readily confused with guano and shadows.

Burrows

Burrows are best censused by counting the number that appear occupied within random or stratified random quadrats or line transects (see Chapter 2, Harris 1984, Harris & Rothery 1988) and then counting the number of burrows that appear occupied. Circular sampling quadrats are easy to use in practice as a fixed length of rope tied to a stake will give a fixed quadrat size. A rope and stake are also easy to carry into the field. Occupied burrows can often be recognised by a range of features such as feathers, excavated earth, droppings, broken eggshells, and smell (especially when young are present). An endoscope (optical fibrescope) can also be used to examine the nest contents, though in practice if the burrow is too long or has too many bends this may be time-consuming or impossible. Digging down to the nest to expose the nest chamber is discouraged. For nocturnal species it may be useful to play recordings of the call to elicit a response (James & Robertson 1985).

One perennial problem with censusing burrow-nesting birds is that it is necessary to distinguish the burrows of different species and to exclude mammal burrows. This is not always straightforward. The simplest way of overcoming this is to survey in places, or at times of year, when only the species under study is present. This will clearly not always be practical, and should endoscopy or playback prove impossible it may be necessary to develop more sophisticated techniques such as that of Alexander & Perrins (1980), which is based on mark–recapture of chicks.

Colonies of ground-nesting species

Many species of seabirds, e.g. gulls, terns, penguins, and albatrosses, nest in colonies on the ground. If the colony is small (less than 200 pairs) and can be easily viewed then the number of nests may be counted directly. For larger colonies it is probably sensible to subdivide the colony and count each section separately. Old nests, determined readily by the lack of a white coat of faeces, should not be included in the counts.

Counts should be carried out at the time of year when adults are most likely to be on the nest (usually from mid-incubation to soon after hatching) and during the time of day when attendance is most stable. This will probably vary between species and colonies, but as a general rule avoid early morning and evenings.

Particularly extensive colonies are probably best censused using line transects or quadrats (see Chapter 2, Thompson & Rothery 1991). To do this first map the colony boundaries and calculate the overall area of the colony. When using transects, define their location, mark them on the gound with string, walk their length, and count all occupied nests up to a set distance (e.g. 1 m) from the transect. Do not count the same nest twice. Alternatively, if using quadrats, locate them at random within the colony, or at equal distances along a transect, and count all occupied nests within each quadrat. It is simple to calculate colony size from the total area of the colony and the total number of occupied nests in, and area of, all transects or quadrats.

Any technique which forces adults away from their nests is traumatic for both birds and observer. It is essential to ensure that disturbance is kept to a minimum and adults should not be kept off the nest for more than 30 minutes, ideally less. Colonies should not be disturbed when it is very wet, cold, or hot for fear of causing egg and chick losses. In addition heavy rain, fog, or wind probably affect count accuracy (Wanless & Harris 1984). Care should be taken to ensure that chicks do not run off, predators do not take advantage of the disturbance, and eggs and young do not get trampled. Such techniques are probably best not used near public areas.

Rather than walking through colonies to count nests, a more rapid and less invasive technique for small colonies is the flush count. In this all birds in the colony are flushed into the air with a loud noise and all flying birds counted. Using this method colony size can only be estimated if the relationship between flying birds and breeding pairs is known. For Arctic Terns *Sterna paradisaea* three flushed birds equals two breeding pairs (Bullock & Gomersall 1981), though this relationship will vary between species.

Tree colonies

Many herons, egrets, storks, and spoonbills nest in dense colonies in trees, though species from several other groups (e.g. crows and weaver birds) do as well. For those which nest in deciduous trees reasonably early in the year, nests are best counted before the leaves have completely emerged, else the nests will become obscured. Occupied nests can often be identified by the presence of fresh nesting material, droppings in the nest or underneath, incubating or attendant adults, or chicks calling in the nest. Alternatively it may be necessary to use a nest mirror (a mirror on the end of a long telescopic pole) to see into the nest cup. Nest mirrors, however, can be heavy and awkward to use (particularly in high winds), and often require two people to use them, one holding the pole, the other looking at the mirror through binoculars. Many herons, egrets, storks, and spoonbills are sensitive to disturbance, especially early in the breeding season, and for these species it is not a good idea to visit until egg laying has commenced. Even then, extreme care should be taken to ensure minimal disturbance.

Fortunately, many tree-nesting species are highly visible from a distance, and so provided a suitable vantage point can be found, sensible nest counts are reasonably straightforward. Observation from custom-built tower hides is often best. Where a suitable ground-based vantage point cannot be found, aerial counts can be undertaken, finances permitting. Aerial counts of Great Blue Herons *Ardea herodias* recorded 87% of the ground total while aerial photographs recorded 83% (Gibbs *et al.* 1988). Aerial methods were considered less disruptive than ground counts, were precise, and had a high repeatability. The least disruptive method used, however, was a ground count of used nests after the breeding season.

Advantages and disadvantages of counting nests in colonies

The biggest advantage of this method is that counts are undertaken at a time of year when the species is highly clumped and thus can be counted in a very cost-effective manner. At other times of the year these species may well be spread over a very much larger area and are consequently very difficult to census. The disadvantages are that it is only suitable for breeding birds, and that care has to be taken to keep disturbance to a minimum. Aerial photography can help avoid such disturbance, though there are cases of colonies failing as a result of helicopters flying over. The identification of the occupant of nesting burrows can be difficult.

Biases

Colony attendance can vary both diurnally and throughout the breeding season and this must be taken into account. Poor vantage points for counting can lead to unknown biases. Some burrow-nesting species have multiple nests in a single burrow, so care needs to be taken to ensure the population is not underestimated.

Counting leks

Lekking species: the known lekking species, taken from Johnsgaard (1994), are: Sage Grouse *Centrocercus urophasianus*, Blue Grouse *Dendragapus obscurus*, Capercaillie *Tetrao urogallus*, Black-billed Capercaillie *T. parvirostris*, Black Grouse *T. tetrix*, Caucasian Black Grouse *T. mlokosiewiczi*, Ruffed Grouse *Bonasa umbellus*, Greater Prairie-chicken *Tympanuchus cupido*, Lesser Prairie-chicken *T. pallidicinctus*, Sharp-tailed Grouse *T. phasianellus*, Tragopans *Tragopan* spp., Peacock-pheasants *Polypletron* spp., Crested Argus *Rheinardia ocellata*, Great Argus *Argusianus argus*, Blue Peafowl *Pavo cristatus*, Wild Turkey *Meleagris gallopavo*, Ocellated Turkey *Agriocharis ocellata*, Houbara Bustard *Chlamydotis undulata*, Little Bustard *Tetrax tetrax*, Great Bustard *Otis tarda*, Australian Bustard *Ardeotis australis*, Great Indian Bustard *A. nigriceps*, Kori Bustard *A. kori*, Denham's Bustard *Neotis denhami*, Black-bellied Bustard *Eupodotis melanogaster*, Rufous-crested Bustard *E. ruficristata*, Bengal Florican *E. bengalensis*, Lesser Florican *E. indica*, Buff-breasted Sandpiper *Tryngites subruficollis*, Ruff *Philomachus pugnax*, Great Snipe *Gallinago media*, Kakapo *Strigops habroptilus*, Band-tailed Barbthroat *Threnetes ruckeri*, Green (Guy's) Hermit *Phaethornis guy*, Long-tailed Hermit *P. superciliosus*, Reddish Hermit *P. ruber*, Little Hermit *P. longuemareus*, White-tipped Sicklebill *Eutoxeres aquila*, Rufous Saberwing *Campylopterus rufus*, Brown Violet-ear *Colibri delphinae*, Green Violet-ear *C. thalassinus*, White-eared Hummingbird *Hylocharis leucotis*, Blue-throated Goldentail *H. eliciae*, White-bellied Emerald *Amazilia chionogaster*, Violet Saberwing *Campylopterus hemileucurus*, Scaly-breasted Hummingbird *Phaeochroa cuvierii*, White-necked Jacobin *Florisuga mellivora*, Violet-headed Hummingbird *Klais guimeti*, Blue-chested Hummingbird *Amazilia amabilis*, Charming Hummingbird *A. decora*, Cinnamon Hummingbird *A. rutila*, Rufous-tailed Hummingbird *A. tzacatl*, White-tailed Emerald *Elvira chionura*, Copper-headed Emerald *E. cupreiceps*, Snowcap *Microchera albocoronata*, Amethyst-throated Hummingbird *Lampornis amethystinus*, Crimson Topaz *Topaza pella*, Marvelous Spatuletail *Loddigesia mirabilis*, Wine-throated Hummingbird *Atthis ellioti*, Calliope Hummingbird *Stellula calliope*, Broad-tailed

Hummingbird *Selasphorus platycercus*, Trogons *Apaloderma* spp., Superb Lyrebird *Menura novaehollandiae*, Albert's Lyrebird *M. alberti*, Ochre-bellied Flycatcher *Mionectes oleagineus*, McConnell's Flycatcher *M. macconnelli*, Gray-hooded Flycatcher *M. rufiventris*, Speckled Mourner *Laniocera rufescens*, Thrush-like Schiffornis *Schiffornis turdinus*, Black-necked Red-cotinga *Phoenicircus nigricollis*, Guianan Red-cotinga *P. carnifex*, Black-and-gold Cotinga *Tijuca atra*, Dusky Piha *Lipaugus fuscocinereus*, Screaming Piha *L. vociferans*, Rufous Piha *L. unirufus*, Red-ruffed Fruitcrow *Pyroderus scutatus*, Bare-necked Umbrellabird *Cephalopterus glabricollis*, Long-wattled Umbrellabird *C. penduliger*, Amazonian Umbrellabird *C. ornatus*, Calfbird (Capuchinbird) *Perissocephalus tricolor*, Three-wattled Bellbird *Procnias tricarunculata*, White Bellbird *P. alba*, Bearded Bellbird *P. averano*, Bare-throated Bellbird *P. nudicollis*, Guianan Cock-of-the-rock *Rupicola rupicola*, Peruvian Cock-of-the-rock *R. peruviana*, Sharpbill *Oxyruncus cristatus*, Crimson-hooded Manakin *Pipra aureola*, Band-tailed Manakin *P. fasciicauda*, Wire-tailed Manakin *P. filicauda*, Red-capped Manakin *P. mentalis*, Golden-headed Manakin *P. erythrocephala*, Round-tailed Manakin *P. chloromeros*, White-crowned Manakin *P. pipra*, Blue-crowned Manakin *P. coronata*, White-fronted Manakin *P. serena*, Long-tailed Manakin *Chiroxiphia linearis*, Blue-backed Manakin *C. pareola*, Swallow-tailed Manakin *C. caudata*, Golden-winged Manakin *Masius chrysopterus*, Pin-tailed Manakin *Ilicura militaris*, White-throated Manakin *Corapipo gutturalis*, White-ruffed Manakin *C. altera*, White-collared Manakin *Manacus candei*, Orange-collared Manakin *M. aurantiacus*, Golden-collared Manakin *M. vitellinus*, White-bearded Manakin *M. manacus*, Fiery-capped Manakin *Machaeropterus pyrocephalus*, Striped Manakin *M. regulus*, Club-winged Manakin *M. deliciosus*, Grey-headed Piprites *Piprites griseiceps*, Yellow-whiskered Greenbul *Andropadus latirostris*, Bowerbirds *Scenopoeetes, Archboldia, Amblyornis, Prionodura, Sericulus, Ptilonorhynchus, Chlamydera*, Standardwing *Semioptera wallacii*, Black-billed Sicklebill *Epimachus albertisi*, Pale-billed Sicklebill *E. bruijnii*, Superb Bird-of-paradise *Lophorina superba*, Western Parotia *Parotia sefilata*, Carola's Parotia *P. carolae*, Lawes' Parotia *P. lawesii*, Wahnes' Parotia *P. wahnesi*, Magnificent Riflebird *Ptiloris magnificus*, Victoria's Riflebird *P. victoriae*, Paradise Riflebird *P. paradiseus*, Magnificent Bird-of-paradise *Cicinnurus magnificus*, Wilson's Bird-of-paradise *C. respublica*, King Bird-of-paradise *C. regius*, Arfak Astrapia *Astrapia nigra*, Splendid Astrapia *A. splendidissima*, Ribbon-tailed Astrapia *A. mayeri*, Stephanie's Astrapia *A. stephaniae*, Huon Astrapia *A. rothschildi*, King-of-Saxony Bird-of-paradise *Pteridophora alberti*, Twelve-wired Bird-of-paradise *Seleucidis melanoleuca*, Red Bird-of-paradise *Paradisaea rubra*, Lesser Bird-of-paradise *P. minor*, Greater Bird-of-paradise *P. apoda*, Raggiana Bird-of-paradise *P. raggiana*, Goldie's Bird-of-paradise *P. decora*, Emperor Bird-of-paradise *P. guilielmi*, Blue Bird-of-paradise *P. rudolphi*, Village (Green) Indigobird *Vidua chalybeata*, Pin-tailed Whydah *V. macroura*, Eastern Paradise-whydah *V. paradisaea*, Broad-tailed Paradise-whydah *V. obtusa*, Jackson's widowbird *Euplectes jacksoni*

Method

In about 150 species of birds (Johnsgaard 1994) males collect in communal display arenas called leks. These may be attended by males for much of the breeding season and much of the day, although females may attend them only briefly to mate. Peak numbers of males of many lekking species occur just before egg laying and often just after dawn (e.g. Cayford & Walker 1991), or just after dusk in the nocturnally displaying Great Snipe.

There is usually little interchange of males between sites. A single count at the peak time is thus usually sufficient, provided the weather conditions are suitable. Prior to undertaking a full census, however, it is important to undertake counts throughout the day and the season at a small number of sites to determine how lek attendance varies (e.g. Cayford & Walker 1991). The optimal timing for a full census can then be decided. In most cases, counting all males is staightforward as the lek generally covers a small area (see Tables 10 and 11 in Johnsgaard 1994). In a few species, such as the Ruffed Grouse and the Superb Lyrebird, males attend dispersed leks and are often widely separated; counting all males is thus more difficult.

Leks are often on traditional sites, but new leks, especially small ones, may appear. Locating unknown leks can be difficult. In some species, calling males (e.g. Black Grouse) can be heard from some distance giving away the presence of the lek. In other cases the lek arena itself may be obvious even when males are not present (flattened vegetation, droppings, feathers, etc.). Changes in land use may make the habitat surrounding a lek unsuitable and thus lek attendance may decline. This may not represent a population decline; rather the birds may have gone elsewhere. Consequently it is important to count all leks in a reasonably large area to determine changes in population level. Counts of traditional leks only can give a false picture.

Advantages and disadvantages

Most males in an area congregate at the lek and thus can be counted during a single count at the optimal time. Females of lekking species are usually inconspicuous and only visit the lek occasionally and thus cannot be counted reliably at leks.

Biases

A few males may not be present at the lek, even at the optimal time. Young birds are particularly likely to be absent and may lek solitarily. Ruff on migration may form leks in areas in which they do not breed.

Counting roosts

Communally roosting species, particularly waders, many wildfowl, parrots, and some passerines

Method

Many species of birds roost communally either during the night or, among coastal species, at high tide when their feeding grounds are covered. Birds are highly clumped at roosts and thus can be efficiently censused at this time. Many species roost only in the non-breeding season, though some species (e.g. colonial corvids such as the Jackdaw *Corvus monedula*) roost during the breeding seasons as well; in this case males go to roost whilst females incubate.

When roosts are small and easily viewed, birds can be counted at the roost. When they are large or hidden, for example in trees or on rooftops, it is best to count flocks of birds (see Counting flocks, below) entering the roost. This is particularly the case at dusk when flocks of birds coming to roost may be visible against the sky, but quite invisible once on the ground or in the trees.

For some estuarine species where alternative feeding areas such as saltmarsh are available once the mudflats have been covered by the incoming tide, more accurate counts can be obtained during very high tides when all potential feeding sites are covered by the sea.

Roosts can only be counted once located. However, as many roost sites are traditional they are often well known. Unknown roost sites can be located by following the flight paths of flocks of birds as dusk, or high tide, approaches. Some coastal species, however, may roost 50 kilometres inland.

Advantages and disadvantages

This is one of the easiest ways of counting many species that are widely dispersed at other times, and is particularly useful outside the breeding season.

Biases

Some roosts are enormous and contain several million birds (e.g. Starling *Sturnus vulgaris* and Quelea *Quelea quelea*). As when counting flocks (see below) large roosts are probably underestimated, especially if the species is small.

Counting flocks

Flocking species, particularly waders, wildfowl, and some passerines

Method

Where the flock is of no more than a few hundred birds, all can be counted directly from a suitable vantage point through binoculars or a telescope. This is easy with large birds but becomes progressively more difficult with larger numbers and smaller birds at greater distances.

With large numbers of birds, or with mobile flocks, such as those in flight, count in tens, twenties, or even greater numbers rather than counting individual birds. Use landmarks to divide large flocks on the ground into smaller groups. Prater (1979) has shown that observers generally overestimate sizes of small flocks (a few hundred birds) but underestimate sizes of large flocks (a few thousand birds). Rapold *et al.* (1985) also document large observer errors in the estimation of flock size.

Advantages and disadvantages

The method is of most use outside the breeding season. Unlike roosts and leks the locations of flocks need not be traditional. When searching an area systematically for flocks the data obtained can be markedly non-normal in distribution (e.g. many areas with no birds, and thus zero counts, and a few areas with very large counts); such data can present problems during statistical analyses.

Biases

Flock size estimates can be markedly wrong and the degree of error varies between observers. Flocks tend to bunch in the centre so that different parts of the flock covering similar areas will not contain the same number of birds.

Counting migrants

Migrating raptors and storks, some passerines

Methods

Diurnal migrants

Several species of migrants pass through bottlenecks on their migration routes. For example, raptors and storks on migration in Europe and the Middle East concentrate at narrow sea crossings and many of the best-known sites, such as the Straits of Gibraltar and the Bosphorus (Turkey), are routinely counted (Porter & Beaman 1985). During the fall in North America, many hawks pass through migration funnels along the barrier islands off the coast and over certain ridgetops of the Appalachians.

Because these species are often widely dispersed at other times of year, counting at bottlenecks, particularly when there is only a limited number of them, can be an extremely efficient manner of censusing. Though the most complete counts are made over the entire migration period, 80–90% of some raptor species pass through a bottleneck during a two- to three-week period, the dates of which are often well known. Teams of observers position themselves at vantage points (often high ground) 6–8 km apart across the breadth of the flyway, which is about the optimal distance to avoid different teams counting the same birds. Each team consists of one to three observers, including an expert on raptor identification, and one transcriber. Each team counts the number of gliding birds (not those circling in thermals) passing per hour. It is useful if one observer counts to the north, one to the south, and one overhead. Ideally the teams should communicate with radio transmitters to avoid duplicate counting. Where numbers become too great to count, it may be sensible to photograph the passing flocks, project the image onto a screen, and count the dots (Smith 1985). Further information on counting migrant raptors is given in the proceedings of the World Conferences on Birds of Prey and Owls (e.g. Newton & Chancellor 1985, Meyburg & Chancellor 1989, 1994) and in Kerlinger (1989).

Nocturnal migrant songbirds

A large proportion of migrants travel at night and some nocturnal migrant songbirds call to one another to keep in touch. These calls are generally specific to the species, though it is often hard to hear them against a background of other noise. Using sensitive microphones and customised software it is possible to distinguish automatically between species and to count the number of each passing overhead at night (Evans 1994). Currently the technology does not count birds that fly above 1000 m, and, for monitoring purposes, it assumes that a constant proportion of each species calls as they fly; whether or not this is the case is unknown.

An alternative approach to studying nocturnal migrants is to observe them through a telescope as their silhouettes pass across the moon's disc at night (Lowery & Newman 1966, Alerstam 1990). With this 'moonwatching' method birds within the narrow cone of sky between the observer and the moon can be seen and counted. Following some rather complex calculations (Lowery & Newman 1966, Alerstam 1990) migration intensity can be determined. Though the method obviously requires cloud-free weather, the smallest songbirds can be detected at a distance of 2 km with a $20 \times$ telescope.

Radar has been used to determine migration routes as well as to calculate the size of migrating flocks. Though not covered in detail here, useful summaries of the uses of radar in ornithology are given in Eastwood (1967) and Alerstam (1990).

Advantages and disadvantages

A large proportion of the populations of some migrants pass through bottlenecks, thus allowing a cost-efficient method of counting these species. Identification of species can be difficult, particularly for high-flying and nocturnal migrants. Large numbers of well-coordinated personnel are needed to count migrating raptors.

Biases

Weather conditions can cause raptor streams to change position, and thus flight paths can be missed. Differing levels of observer experience can lead to bias through misidentification. Double counting by observers who are spaced out to count over a broad front can lead to overestimates of population size. At midday, some raptors may migrate at heights invisible to the naked eye, and, in general, the uncertainty over the heights at which individual species migrate makes calculations of migration intensity difficult. Some migration bottlenecks may have been overlooked.

Territory mapping

Territorial breeding species: some ducks, gamebirds, and raptors, and most passerines

Method

During the breeding season many species of birds are territorial. Males sing to defend their territories, nests are built within them, and the boundaries between territories are often clearly defined by disputes with neighbouring birds. The breeding territory can thus readily be used as a census unit, and territory mapping, in which all signs of territory occupancy are marked on a large-scale map of a plot, can be used as an effective census tool. The aim of territory mapping is to determine how many territories of each species there are on a given plot. Standardised techniques are given in Kendeigh (1944), Enemar (1959), IBCC (1969), Marchant *et al.* (1990), and Bibby *et al.* (1992).

First, the study plot needs to be mapped at a scale of about 1 : 2500. To enable species' registrations to be located accurately on the map, any obvious features which can be readily identified during each visit (e.g. buildings, ponds, isolated trees, tracks, rides, hedges, etc.) should be marked on the map. The size of the plot should be such that a reasonable number of territories of each species are present. A realistic goal would be to ensure that half of the species in the plot are represented by five or more territories, the other half by fewer (Terborgh 1989). Though these numbers may be minimal for statistical purposes, they require a great deal of fieldwork. Plot size will vary with habitat as bird density, diversity, and conspicuousness vary with habitat; 15–20 ha in closed habitats such as temperate woodland would be suitable (though perhaps half this in tropical forest), with 60–80 ha in more open habitats (e.g. agricultural, moorland, grassland, and steppe). If the census is being undertaken for long-term population monitoring, it may be important to ensure that the habitat is not successional.

Long, thin plots are unsuitable for territory mapping because the ratio of edge to area is high and many bird territories will overlap the plot boundary. Territories along the edge of the plot cause problems because it is often difficult to determine whether or not a particular territory 'belongs' to the plot. Round or square plots are preferable. In addition try to avoid using a species-rich feature of the landscape (e.g. a hedge) as a plot boundary, as this will serve to exacerbate any edge problems.

Several visits need to be made to each plot during the breeding season. In temperate regions, where breeding seasons are clearly defined, 5–10 visits per plot is suitable; open habitats would be at the bottom end of this range, while woodland and forest with high densities of birds would be at the top. The visits should be spread throughout the season to ensure both early- and late-breeding species are included in the censuses. In tropical areas with less clearly defined seasons, number and timing of visits need careful consideration.

Many birds sing most during the first hour after dawn. As a consequence this period can be confusing in areas with high densities of birds, and is probably best avoided. Surveys should be completed by midday, as many species sing less in the afternoon. Temperate forest can be surveyed at a rate of about 5 ha per hour, tropical forests at about half this rate, and more open habitats at about 20 ha per hour. Each visit to a plot can thus be undertaken during a morning.

Prior to the start of the season you will need to produce several copies of the plot map, one for each field visit and, ultimately, one for each species' map (see below). Large maps can be awkward to use in the field, and are best attached to a clipboard. This can be covered by a large polythene bag if it is raining. The plot should be covered at a slow walking pace with the route approaching within 50 m of every point on the plot. Each bird encountered is marked on the map using standard codes (Box 8.1). Evidence of nesting, such as nests, alarm calling, and birds carrying nesting material or food, are particularly useful, as are simultaneous observations of different individuals of the same species (e.g. counter-singing or fighting males). Without these the subsequent analysis of the maps (see below) is much less accurate (Tomiałojć 1980). It is necessary to work slowly and carefully to build up these records and to record inconspicuous species, though covering the plot too slowly may lead to unintentional double counts of the same individual. Mapping should extend slightly outside the study area to ensure that the territory boundaries of species at the edge of the plot are recorded (see Analysis of maps, below).

The territory mapping method can be extended to cover a much larger geographical area simply by ignoring common species and mapping at a much bigger scale (e.g. 1 : 10 000, Robertson & Skoglund 1985). By doing this, species which range over a much larger area but which are nevertheless territorial (e.g. raptors) can be censused.

Analysis of maps

At the end of the season all the information from the individual visit maps are transferred to species' maps (one per species). Registrations transferred from the first-visit map are denoted by the letter A, the second by B, and so on. All the records of a particular species from the first visit are transferred to its species map, but with A replacing the species code. The symbols from Box 8.1 are also incorporated on the species map. This is repeated for each of the visits until a map containing all of the registrations from all of the visits is produced for each species.

The symbols on the species' maps should form clusters around which non-overlapping rings representing approximate territory boundaries can be drawn. Conventionally, at least two registrations are needed to define a cluster if there were 5–7 visits or three if 8–10 visits. To avoid including temporary migrants, records in the cluster must be from at least ten days apart. Simultaneous registrations indicate different individuals and should never be incorporated into the same cluster unless they are thought to be two adults of a pair. Records of nests can be counted as a cluster even in the absence of sufficient records of the adults.

Dealing with edge clusters, those that overlap the plot boundary, is problematical, and several analytical methods have been used. Treat all edge territories as belonging to the plot; include them if more than half of the registrations within the cluster lie in the plot; or use the proportion of a cluster's registrations that lie within the plot to calculate a fraction of a territory. The first method should not be used to estimate densities as it will lead to overestimates.

Box 8.1. **Activity codes for use in mapping censuses in Finland**

These activity codes have been developed from, and are very similar to, the mapping codes used by the British Trust for Ornithology. Most examples are for the chaffinch *Fringilla coelebs*. Some countries have standard codes for each species name (e.g. in the UK, CH = chaffinch).

A chaffinch in song.

A chaffinch in song (exact location shown by the point).

A chaffinch in song (location is not exact; the point where the observation was made is shown by the cross).

$F_{coe}\male$ A male chaffinch repeatedly giving alarm calls or other vocalisations (not song) thought to have strong territorial significance.

$F_{coe}\male$ A male chaffinch calling.

$F_{coe}\male$, $F_{coe}\female$, F_{coe}, $F_{coe}\,2\male\,1\female$, $3\,F_{coe}\,juv$

Chaffinch sight records, with age, sex, or number of birds if appropriate. Use F_{coe} $\male\female$ to indicate one pair of chaffinches, i.e. $2\,F_{coe}\,\male\female$ means two pairs together.

$F_{coe}\male^{f}$ A male chaffinch carrying food (or faeces).

$F_{coe}\female^{m}$ A female chaffinch carrying nest material.

$F_{coe}^{*\,2E3N}$ An occupied nest of chaffinches, with 2 eggs (E) and 3 nestlings (N); * shows the location. Do not mark unoccupied nests, which are not of territorial significance by themselves.

$P_{maj}^{\boxminus\,10E}$ Great Tit *Parus major* nesting in a specially provided site. Please remember to use this special symbol for a nest in a nest box.

F_{coe}^{*P} Chaffinch nest with a parent bird incubating or warming young.

$F_{coe}\,juv$ A chaffinch fledgling.

-$F_{coe}\,fam$ Juvenile chaffinches with parent(s) in attendance.

Movements of birds can be indicated by an arrow using the following conventions:

$F_{coe}\male$ A calling male chaffinch flying over (seen only in flight).

$F_{coe}\,\female$ A female chaffinch moving between perches. The solid line indicates it was
$F_{coe}\,\female$ definitely the same bird.

A singing chaffinch perched, then flying away (not seen to land).

A male chaffinch flying in and landing (first seen in flight).

 A Siskin *Carduelis spinus* circling above the forest.

The following conventions indicate which registrations relate to different, and which to the same, individual birds. Their proper use will be essential for the accurate assessment of clusters.

 Two chaffinches in song at the same time, i.e. definitely different birds. The hatched line indicates a simultaneous sighting/hearing of song and is of great value in separating territories.

 The solid line indicates that the registrations definitely refer to the same bird.

The question-marked solid line indicates that the sightings/songs probably relate to the same bird. This convention is of particular use when your census route brings you back past an area already covered – it is possible to mark new positions of (probably the same) birds recorded before, without risk of double recording. If you record birds without using the question-marked solid line, over-estimation of territories will result.

 No line joining the registrations – there is no assumption as to whether the records concern different birds, but depending on the pattern of other registrations they may be treated as if only one bird was involved (a question-marked dotted line indicates that the sightings/songs were almost certainly of different birds).

 Two chaffinch nests occupied simultaneously, and thus belonging to different pairs. Only adjacent nests need to be marked in this way. Where they are marked without a line, it will be assumed that they were first and second broods, or a replacement nest following an earlier failure.

 An aggressive encounter between two chaffinches; may be accompanied by notes on vocalisations.

Clusters can be difficult to differentiate and may overlap. For inconspicuous species there may be few registrations per cluster and no simultaneous registrations. Thus, despite the existence of standard guidelines (e.g. IBCC 1969, Marchant 1983, Marchant *et al.* 1990, Bibby *et al.* 1992) analysis of species' maps can be subjective and requires experience, as well as time.

Though map analysis is generally undertaken at the end of the season, if it were undertaken during the season, fieldwork could be targeted at clarifying confusing situations. The species' maps could even be taken into the field.

Advantages and disadvantages

Territory mapping is very time-consuming and thus expensive. It is not suitable for species which are colonial, which live in loose groups, or whose territories are large relative to the study area. It can only be used when birds are territorial, and thus is largely only suitable for breeding birds, though there is growing evidence of winter territoriality of individuals, rather than pairs, among migrant species on their wintering grounds (e.g. Schwartz 1964, Rappole & Warner 1980, Kelsey 1989). It does, however, yield a map of bird distributions which can be particularly useful for analysing bird–habitat associations. Because of the great amount of time spent in the field, the method is better 'buffered' against environmental variation (e.g. weather and timing of visits in relation to a species' breeding cycle) than other less time-consuming techniques, such as point count and line transect. It also allows a relatively straightforward calculation of densities.

Biases

In some species unpaired birds sing more (see above), and it is unclear whether it is the breeding population that is being censused. In a study of Ovenbirds and Kentucky Warblers all unpaired males were detected during territory mapping but only 50% of paired male Ovenbirds and 65% of paired male Kentucky Warblers were (Gibbs & Wenny 1993).

The method assumes that birds live in pairs in fixed, discrete, and non-overlapping ranges, and this is often not the case (e.g. polygynous species, polyterritorial species). Despite standard guidelines for map interpretation, there is nevertheless a good degree of subjectivity involved, and this can lead to inter-observer variation. The method can be unreliable at high densities, if birds are not readily visible, if registrations are plotted innacurately, or if it is difficult to obtain many simultaneous registrations.

Point counts

Highly visible or vocal species, often passerines, in a wide variety of habitats

Method

A point count is a count undertaken from a fixed location for a fixed time period. It can be undertaken at any time of year, and is not restricted to the breeding season. Point counts can be used to provide estimates of the relative abundance of each species or, if coupled with distance estimation, can yield absolute densities, provided some important assumptions are not broken (Reynolds *et al.* 1980, Buckland 1987).

Point-count stations (the position from which the count is made) should be laid out within the study plot either in a systematic manner (e.g. on a grid) or in a random manner, stratified or not. The stations should not be too close together, as some individuals would be counted at more than one counting station, which could spuriously inflate the sample size and influence the precision of the results. A sensible minimum distance is 200 m. If the distance between points is too great, however, too much time will be wasted travelling between the counting stations. As a reasonably large number of point counts (more than 20) will be needed from each study plot, point counting is not a suitable technique for small study areas. Twenty counts can readily be made in a morning starting soon after dawn.

Wait for a few minutes before beginning to count at each station; this allows the birds to settle down following the observer's arrival. Count for a fixed amount of time at each station. Ideally, this should be somewhere between three and ten minutes, the actual duration depending on habitat and the bird communities present. If counts are too short, individuals are likely to be overlooked, while if they are too long, some birds may be counted twice. A study of count duration in six habitats in Britain suggested that counts did not need to be longer than ten minutes, and that five minutes was usually adequate (Fuller & Langslow 1984). Record all birds seen or heard. Endeavour to count each individual only once. Most registrations will occur in the first few minutes; thus counting for too long can be inefficient. The time saved from counting for a shorter period can be used to count at more points or to cut down the total time spent in the field. In areas with a very rich bird fauna or where species are hard to detect or identify, for example in a tropical rainforest, it may be necessary to count for longer than ten minutes.

In habitats with high densities of birds it is easy to confuse different individuals, or to be uncertain whether you have already recorded a particular individual or not. A simple way of resolving this is to record their approximate positions in a page of a notebook. This can be divided into four quarters, and birds recorded in these quarters (e.g. left and to the front, right and behind, etc.) marked accordingly by a species code. If you are counting in several different distance bands (see below), these could also be drawn as concentric circles around your central position.

If all that is required are relative indices of abundance then count up to an unlimited distance or only within an arbitrary range, such as 25 m from the observer. However, such indices are liable to be uncomparable between species owing to differences in detectability between them, and are liable to differ between habitats, because 'unlimited' distance is much closer in a closed habitat (e.g. woodland) than in an open one. Bird detectability decreases with distance from the observer, and to use point counts to estimate density a few assumptions are necessary. These are that all birds are detected at the centre of the count area, and that detectability falls off with distance in a known manner (see Chapter 2). In practice there are three ways of incorporating distance estimation into point counts to enable density to be calculated. The simplest, and recommended, method is to have two counting bands, and to record birds up to a fixed distance (e.g. 25 m in a dense forest or 50 m in a more open habitat) and beyond that distance separately. Simple formulae (based on the possible manner in which detectability falls off with distance) are then used to calculate the density of each species (see Chapter 2, Bibby *et al.* 1985, 1992, Buckland 1987, Buckland *et al.* 1993).

Increasingly complex methods require using either several distance bands rather than just two, or attempting to estimate the distance to every contact (Buckland 1987). Try to ensure that the first distance band is not so close that no birds are recorded in it, else density calculations will prove impossible.

It is often advisable to undertake at least two separate counts at each counting station, one in the first half of the season and one in the second. This not only will ensure that both early and late breeders are recorded during the counts, but will, in part, take into account seasonal variation in the detectability, since a species, though present, may be more detectable during one part of the season than another. In general, the maximum value for each species at each counting station should be used in analyses of density. This increases the chance that all birds are detected at the centre of the count area, which is an important assumption in density estimation (see below). The maximum value need not be used if only relative indices are required; rather a mean value can be used. If several counts are made at each counting station it can sometimes be difficult to relocate the precise counting station on subsequent visits. It is thus necessary to mark their location in a reasonably obvious manner (e.g. brightly coloured tape wrapped around a post or vegetation), particularly in habitats in which the vegetation is likely to grow rapidly between visits.

As well as varying seasonally, detectability can vary through the day. This can be assessed by plotting the number of contacts against time of day and then calculating the moving average for a set time period (e.g. 60 minutes). The highest average is rescaled to a value of one and detectability at all other time periods can be determined by comparison with this maximum value (Palmeirim & Rabaça 1993). Where point counts are spread out over a morning or a day, it is then possible to correct all to a standard time of day. Where several species are being monitored, however, it would be necessary to undertake such analyses for all species.

The North American Breeding Bird Survey (Robbins *et al.* 1986) uses 50 three-minute counts at intervals of 0.8 km along randomly selected roadside routes. Birds are counted at the height of the breeding season, starting 30 minutes after sunrise, and all birds heard and seen with 0.4 km of the road are recorded at each stop. Each route takes about 4–4$\frac{1}{2}$ hours. No distance estimation is involved; thus all species are measured by a relative index. Though it has gathered an enormous range of information (2 000 routes are counted each year and used to monitor about 230 species), the roadside nature of the scheme has led to problems in interpretation since habitat change along roads is unlikely to be representative of habitat change throughout America as a whole. Such potential problems should always be considered if the census technique is to be used as the basis of a long-term monitoring scheme.

Advantages and disadvantages

Point counts are widely used to census songbirds, but are of little use for less detectable species. However, point counts have been used to census waders, waterfowl, and nocturnal birds in Finland (Koskimies & Väisänen 1991). Because most birds are detected by song, a high level of observer experience is required. Counting stations are relatively easy to allocate randomly, which is not always the case for territory-mapping plots or transects. Point counts are more suitable than transects where habitat is patchy, though much less so in open habitats where birds are likely to flee from the observer. Point counts are unsuitable for species which are easily disturbed. They are, however, very efficient for gathering large amounts of data quickly. Point counts can be used outside the breeding season.

Biases

Estimation of density assumes that all birds at the centre of the count area (i.e. where the observer stands) are recorded. This will not be the case if birds flee from the observer, nor if the species is particularly skulking. As the area sampled by a point count increases geometrically with distance from the observer, small errors in detecting birds close to the observer can seriously bias density estimates (Verner 1985). Biases will similarly occur if birds are attracted to the observer. Where birds are highly mobile the same bird may be recorded twice. Because of the short length of time spent in the field, point counts can be markedly influenced by weather conditions. Counts should thus not be carried out in stong winds, rain, or cold weather.

Line transects

Birds of extensive open habitats, e.g. shrub–steppe and moorland, offshore seabirds, and waterbirds

Method

Line transects are undertaken by observers moving along a fixed route and recording the birds they see on either side of the route. Transects can be walked (or driven) on land, sailed on the sea, or flown in the air. Because the observer needs to be able to move freely through the land, sea, or air, transects are most suitable for large areas of continuous, open habitat.

First, the transect route(s) need to be chosen. Assuring that their location is as random as possible is crucial to the success of the scheme. If, for example, a route were to follow a path, a hedge, a stream, or a road, the results obtained could be markedly biased by the influence that these linear features might have on bird populations. A similar example would be of transect counts of fish-eating seabirds made from fishing trawlers which are, like the seabirds, actively seeking out fish stocks. The location of such transects could not be considered to be random, and would bias (probably upwardly inflate) any estimates made from the counts. The difficulty in randomly allocating routes because of access problems is one of the biggest disadvantages of the transect method. The Breeding Bird Survey in the United Kingdom (Marchant 1994) uses transects located on a north–south axis within randomly allocated 1-km × 1-km grid squares. A transect route could even be square or rectangular allowing the observer to end up at the starting point. Where maps are insufficient to plan a route precisely, it is a good idea to walk along compass bearings (Koskimies & Väisänen 1991). In principle a transect could be circular, though it might be difficult to follow such a route in practice, and even harder to use exactly the same route on subsequent visits. In addition if counting were undertaken in distance bands and the circumference of the circle were too small, the area of the internal band would be greater than that of the external band, which would add complications at the analysis stage.

The total length of transect route will vary depending upon the study in question. Practical considerations (e.g. time available to spend in the field, and size of the area to be censused) may well be overwhelming considerations. Ideally, split the total length down into several shorter lengths. These could either continue one on from the other or be wholly independent of one another. The latter may be more useful for analytical purposes as the separate lengths of the transect can be considered statistically independent, provided they are sufficiently well separated. In Finland, breeding birds are censused along rectangular transects with a total length of about 5 km (Koskimies & Väisänen 1991). In the United Kingdom, two transects, one south to north, and the other north to south, each of 1 km in length are walked in each 1-km square.

If several different transects are to be undertaken on a plot, they should be sufficiently far apart to ensure that the same individual is not counted on different transects. Sensible distances might be 150–200 m in closed habitats, but 250–500 m in open habitats. The distance between transects should be greater in open habitats because birds are more visible over greater distances, and because birds are more likely to flee greater distances from an observer.

Once the transect routes have been planned it is then necessary to decide how many visits are to be made to each route, and up to what distance(s) from the transect you intend to count. As for point counts it may be sensible to repeat each transect one or more times to maximise the chance of recording all species, since detectability varies seasonally (either because species are absent or simply unobtrusive).

Methods for estimating density are very similar to those used in point counts (see Chapter 2 for details). Simple indices of the number of birds recorded per unit length of transect can be obtained by counting birds either up to an unlimited distance, or to a single fixed distance (a fixed strip transect), on either side of the transect. Approximate densities can be calculated from the fixed strip transect but rely on the assumption that all birds are detected, which is unlikely to be the case for all but the narrowest of strips and the most detectable of species. More reliable measures of density, however, can be obtained from a fixed-distance line transect in which birds are recorded separately in a central zone (the 'near belt') and beyond it (the 'far belt'). The width of the near belt should be such that about half of all records fall within it and half beyond. Its width will be greater in more open habitats and in areas with less dense bird populations. The near belt is commonly placed 25 m either side of the transect line. Density can then be calculated from relatively simple formulae which are based on the manner in which detectability falls off with distance. Examples are given in Järvinen & Väisänen (1975), Burnham *et al.* (1980), Bibby *et al.* (1992), and Buckland *et al.* (1993). Increasingly complex methods involve recording in several different bands (Emlen 1977), and measuring the perpendicular distance from the transect route to every record (Burnham *et al.* 1980).

In practice it may be sensible to map all bird records onto a schematic representation of the transect in your notebook or onto a recording sheet. It might also help if distance bands were drawn onto this. Recording the birds in this manner means that a variety of different techniques can ultimately be used to analyse the data. If birds are being recorded in separate distance bands, check that the distances can be reliably estimated. Optical range finders are available and could be used. Try to standardise the rate of movement along the transect route; walking too fast misses birds, but walking too slowly may result in double counting. A walking rate of 2 km per hour is reasonable in open habitats, though 1 km per hour would be more realistic in forest. Record the location where the bird was first observed, rather than where it fled to. Record birds flying straight down, or singing above, the central belt as being within the central belt, but record birds flying over as being in the far belt.

Transects at sea

Away from their breeding colonies seabirds are frequently censused by transect from a ship. Seabirds present particular problems because they are often recorded in flight, and their speed of flight in relation to the speed of the ship through the water, and their direction of flight relative to that of the ship, can influence the results very markedly. In particular the same bird may be recorded repeatedly.

Tasker *et al.* (1984) have proposed a simple, standard method for transects at sea. This is enlarged upon in a very useful technical manual by Komdeur *et al.* (1992). The method is as follows. The observer counts looking out from one side of the ship only. Birds on the surface of the sea and flying birds are recorded using different methods, though both only include birds up to a perpendicular distance of 300 m from the ship. For birds on the sea, all birds passed in a 10-minute period are noted. To calculate the area of the transect the ship's speed is recorded. As an example, a ship travelling at 10 knots moves 3.2 km in 10 minutes; thus its transect area would be 0.96 km^2. Flying birds are counted during the same 10-minute period in several instantaneous 'snapshots' 300 m wide (the perpendicular distance out from the ship) and of a length as far ahead as the observer thinks all birds are visible. Thus if the observer thinks that for the species in question, and under the prevailing conditions, birds can only be recorded up to 400 m, then eight instantaneous counts, each looking 400 m ahead, and each separated by 1.25 minutes, would be needed during the 3.2-km transect. A watch with a beeper can be used to tell the observer when to count. The purpose of the 'snapshot' method is to obtain a record at a precise instant of the birds present in discrete sections of the transect; this is like taking a photograph of all flying birds occurring in sections along the transect. Counts of flying birds and birds on the sea can be summed to calculate an overall density. For some species, the 300-m band may be too far, and this distance may have to be reduced to 150–175 m. Birds 'associated' with the ship (e.g. following it) should either be ignored, or recorded separately.

Transects from the air

Waterfowl and seabirds are sometimes counted from the air while flying along transects of a known length and width (Komdeur *et al.* 1992). Though the use of a plane can be expensive, the speed of the plane, compared to that of a ship, does mean that the chances of double counting the same birds, and thus overestimating density, are reduced.

The width of the transect will vary with the particular application, but an overall width of about 200 m (100 m on either side of the plane) is sensible. From the air it can often be difficult to determine whether or not a particular bird is within the transect, and it is helpful to mark the aeroplane windows in such a way that you can readily determine from your seat where the edge of the transect lies. The simplest way to do this is to mark out the transect distance on the ground with three poles (two at the extremities, and one in the middle), to fly over the middle pole at the set survey altitude, and to mark the positions of the poles at either side of the transect on to the aeroplane windows. Though two observers, one to look out to each side of the plane, are ideal, this is not strictly necessary. Under some conditions, for example owing to glare from the sun, it may simply not be possible. The plane should be flown at a fixed and reasonably low (e.g. 50–100-m) altitude. Identification becomes difficult if the plane is too high, and the birds pass below too swiftly if it is too low. The transect should be flown at the lowest safe speed (e.g. 150 km per hour). In practice this will often still be too fast to count every individual bird, and quick estimates of flock size are often needed instead. Bird density can be calculated from the number of birds counted and the overall area of the transect (from its width, the speed of the plane, and the time taken to complete the transect).

Advantages and disadvantages

Transects can be undertaken at any time of year, on land, on sea, or in the air. They are suited to large areas of homogeneous habitat, and are particularly useful where bird populations occur at low density. Estimates of density can be calculated. The area sampled by a line transect increases linearly away from the transect line; thus errors in detecting birds close to the observer and in distance estimation are less likely to bias density estimation than in point counts. Random allocation of transect routes can be particularly difficult in some habitats. Because the observer is continually on the move, identification can also be difficult. The high costs of transects at sea can be reduced by observing from ships involved in other activities (though this may introduce some biases). Transects from the air are sometimes too quick to allow precise counts and the identification of some species, their age, and sex.

Biases

Density can only be estimated on the assumption that birds on the transect line are not missed (e.g. because one was walking too fast), that birds do not move before being detected (e.g. if they are disturbed by the observer), that they are not counted twice (because one was walking too slowly), that distance is estimated without error, and that all observations are independent events (e.g. one bird is not detected because of the alarm calls of another). In practice many of these assumptions will not be met and all may lead to bias.

Response to playback
Many species that cannot easily be seen

Method

Some species of birds are notoriously difficult to see, but will respond to a tape-recording of their song or call. Recordings of the songs and calls of many species are now available commercially and can be copied to tape. Ideally, use a tape loop, so the song will continue to be broadcast until the recorder is switched off. If a tape loop is not available, start recording at the beginning of the tape, so that it can easily be wound back to the start of the recording. The song can be broadcast from a hand-held loudspeaker or from one mounted on a vehicle. One problem with the use of playbacks is that some individuals or species may habituate (cease to respond) to the playback if it is used too frequently. Playback can be used alongside other census methods, for example during territory mapping or line transects.

In a study of the Burrowing Owl *Speotyto cunicularia* 53% more individuals were recorded along a transect when calls were broadcast than when they were not (Haug & Didiuk 1993). Similarly, for a range of North American raptors (Cooper's Hawk *Accipiter cooperi*, Red-shouldered Hawk *Buteo lineatus*, Broad-winged Hawk *B. platypterus*, and Barred Owl *Strix varia*) the contact rates for each species were significantly greater when using playback (Mosher *et al.* 1990).

Advantages and disadvantages

Skulking, secretive, and nocturnal species that would otherwise be overlooked can be located and censused. Some species may habituate to the playbacks.

Biases

Care should be taken to ensure that playbacks are broadcast for set durations at a standard volume under set conditions (e.g. time of day), else the responses will vary. Ensure that the use of playbacks is noted, otherwise it may not be possible to repeat the survey precisely. This has particularly been the case when tape lures have been used in mist netting; sometimes tapes have been used, sometimes not, and their use has often gone unrecorded.

Catch per unit effort (mist netting)

Mainly passerines, particularly of woodland and scrub, but also some riverine, dense undergrowth, and canopy species

Method

By placing standard lengths and types of mist nets in standard locations, for standard time periods under similar conditions, this method can be used to monitor changes in population level, productivity, and survival. Several schemes use capture per unit effort as a monitoring tool. The best known are the Constant Effort Sites (CES) scheme in the UK (e.g. Peach & Baillie 1991), the Mettnau–Reitz–Illmitz (MRI) scheme in Germany and Austria (Berthold *et al.* 1986), and the Monitoring Avian Productivity and Survival (MAPS) scheme in the USA (DeSante *et al.* 1993). These schemes differ from all the methods discussed so far since demographic, as well as long-term population information, are collected. Thus rather than simply documenting year on year changes in population level, these methods can help interpret such changes by highlighting whether productivity or survival are possible causes of the population changes. Only qualified ringers (banders) can use this method.

The MAPS program provides very specific methodological detail for participants (DeSante *et al.* 1993), much of which was developed from the CES methods. Sites are chosen to fulfil the following requirements: they are at least 9 ha in size (though preferably up to 20 ha), contain a reasonable avian breeding population, are away from areas where transient and migrant birds tend to congregate, and are in areas where active habitat management ensures that the habitat is held in lower successional stages. The number and length of nets set will depend on the workforce available, though these must remain constant from year to year. Where one or two people are working the site, it is recommended that ten 12-m, 30-mm-mesh, four-tier, black, tethered, nylon mist nets are set uniformly within a 7–8-ha netting area within the whole plot, thus giving a density of about 1.25–1.5 nets per ha. Nets are placed where capture efficiency is likely to be maximised, for example near water, along rides, or at the woodland edge. Nets must be set in exactly the same position each year. The nets are operated for six morning hours per day (beginning at local sunrise), for 12 days, each about 10 days apart, from early May to late August. Netting should only start when most spring migrants have already passed through, and should stop when autumn migrants begin to appear. No lures or playbacks should be used to attract birds to the nets. All birds caught, including retraps, should be identified, aged, and sexed (see e.g. Pyle *et al.* 1987, Svensson 1992, Baker 1993) and all unringed birds should be ringed.

It is essential to standardise catching time and the number of nets in a site. Simply calculating birds per 10 m of net per hour is not sufficient, because doubling the number of nets in a site will not double the number of birds caught, as each can be caught just once. For the same reason, catching for twice as long will not double the number of birds caught, particularly as this would usually mean that more netting was done outside the period of peak activity.

Care must even be taken to standardise the mesh of the nets. Species below 16 g are caught more frequently in 30-mm- than 36-mm-mesh mist nets, while the reverse is true for those above 26 g (Pardieck & Waide 1992). Thus studies should not compare captures made with different mesh sizes. To help stop this happening, mesh size should be specified in publications.

Constant-effort ringing is used largely to study birds of woodland and scrub, but is also good for surveying skulking rainforest undergrowth species. It has also been used to monitor riverine species (Dipper *Cinclus cinclus*, Kingfisher *Alcedo atthis* and Grey Wagtail *Motacilla cinerea*) and was shown to correlate well with actual abundance of these species (Ormerod *et al.* 1988). It can also be used for canopy species, though in this case nets need to be raised many metres above the ground using pulleys or telescopic aluminium poles (Meyers and Pardieck 1993). A gunsling can be used for firing a line up to 45 m in the canopy and then using a pulley to pull up a mist net (Munn 1993).

The information obtained from such schemes has several uses. First, changes in the size of the adult population can be calculated. This can either be a simple index based on the number of adult birds caught, in which between-year comparisons are based on the number of individuals caught during the season irrespective of how many times each was caught, or it can be absolute population levels calculated from capture–recapture methods (see below and Chapter 2). Second, indices of post-fledging productivity can be calculated from the ratio of juveniles to adults caught late in the breeding season. Finally adult survival rates can be calculated from between-year retraps of ringed birds (Buckland & Baillie 1987, Peach *et al.* 1990). Because capture probability does not vary between years, survival-rate analyses are greatly simplified by constant-effort ringing. Such estimates are bound to be minimum because adults that survived between years, but did not return to the same site, are assumed to have died. Increasingly sophisticated methods have been developed to calculate survival rates (e.g. Pollock *et al.* 1990, Lebreton *et al.* 1992, and the SURGE routines of Clobert *et al.* 1987). First-year survivorship (survival rates of young birds) cannot be so readily calculated as they often do not return to their natal site to breed.

Advantages and disadvantages

Unlike most other census methods, capture per unit effort provides information on productivity and survival. Constant-effort schemes are an excellent way of directing the efforts of numerous ringers that would otherwise ring in non-standardised and less useful ways. Long periods of training followed by application for a license are necessary before any ringing can be undertaken in some countries. This makes it an unattractive method for many. In addition it is time-consuming, sites are often chosen rather than randomly allocated, and habitat succession at sites can confuse the long-term picture. As a consequence it is not the most appropriate method for monitoring population levels. It can, however, be useful for censusing species that live in habitats within which observation is difficult (e.g. dense undergrowth, forest canopy, and reed-bed). The method is best for species with high retrap probabilities (e.g. warblers).

Biases

Because of the constraints on finding somewhere suitable to constant-effort ring, sites are rarely randomly allocated and thus between-year changes may not faithfully represent changes over a larger, e.g. national, scale. Any change in methods between years, for example change in location, number, and length of nets and their mesh size, could lead to bias, as could successional changes in the habitat. Some individuals of some species may become 'trap-shy' and thus will actively avoid being recaught.

Mark–release–recapture
A wide range of species

Method

Birds are caught and individually marked, and from the fraction recaptured it is possible to estimate population size using a variety of different analytical techniques. If the marks can be seen from a distance then recapture may not be necessary. Though colour rings are the most usual method of marking, several other techniques, e.g. wing tags, neck collars, and radio transmitters, are available (see e.g. Bibby *et al.* 1992).

Full details of the methods and assumptions of mark–release–recapture are given in Chapter 2.

Catching birds and marking them requires considerable training, and in many countries requires a license. The British Trust for Ornithology's *Ringer's Manual* (BTO 1984) gives a good review of the techniques for capturing and ringing and is available to members of ringing schemes.

Advantages and disadvantages

In practice, mark–release–recapture is rarely used to estimate population size of birds. For most species it is hard work to catch a high enough sample and there are many sources of error. As most species of bird are readily observable, other techniques are usually preferred.

Biases

All methods require that numerous assumptions are not broken (see Chapter 2); this is rarely the case in practice.

Dropping counts

Wildfowl and gamebirds

Method

Determining the distribution of feeding wildfowl can be very time-consuming because they often feed in flocks which frequently move between sites. In some instances daily observations are needed to determine which sites the flocks visit. A simple way of overcoming this is to count the density of their droppings at each site (Owen 1971). A single count can give a relative measure, but a much more accurate measure is obtained if plots are cleared on a regular basis.

In order to do this it is first necessary to determine how long the droppings last before they disintegrate and become either indistinguishable one from another or completely unidentifiable. Fresh droppings can be marked with bamboo stakes and then revisited over a period of days to determine how long they last prior to disintegration. Inevitably, they will disintegrate much more quickly in the rain and when subject to trampling. Bamboo stakes are then placed at random locations (20 gives a good sample if the species is reasonably common) throughout the feeding site. All droppings within a given radius of each bamboo stake are then removed. The simplest way to measure this radius in the field is to use a length of string tied to the bamboo stake. If a spoon is tied to the other end, it can be used to 'flick' the droppings off the circular plot.

The area is then revisited at an interval such that all droppings produced in the interim period will still be visible, and the numbers present in the set radius around each of the randomly allocated stakes counted. The mean number of droppings produced per unit area per day can then be calculated from the number of droppings counted, the number of days between clearing and counting, and the area of each circular sample plot. The whole procedure can then be repeated if required.

These data provide only a relative measure of the extent to which different sites are used, though they can be converted to the number of bird-days by estimating the dropping rate of the species. This technique involves watching an individual bird's bottom for a period of 10 or 15 minutes. If the bird turns out of sight or the view becomes blocked, switch observations to another individual. Intake rate may vary with position in the flock; thus to determine a mean dropping rate, observations should be made of birds throughout the flock.

Gamebirds also have persistent and recognisable droppings. This is particularly useful for surveying elusive forest pheasants. Presence/absence of droppings during a timed search or the number of droppings found along transects are frequently used methods.

Advantages and disadvantages

This is a very useful way of censusing elusive forest pheasants and determining site usage by wildfowl remotely. Some sites may be visited by several species, and distinguishing between the droppings may be difficult.

Biases

Heavy rain and trampling can cause droppings to disintegrate and make counting more difficult.

Timed species counts

High-diversity communities, particularly tropical forests. Also birds of savannah and semi-arid areas (all groups)

Method

This method yields relative indices of abundance and is based on the simplistic assumption that, when one is birdwatching, common birds are on average the first to be noted, while rare birds usually take longer to find. The time to first observation is thus a crude measure of abundance, and can be used to make comparisons both between and within species.

Walk slowly through the study area for a set period (e.g. an hour) and record the time at which each species was first seen. Subsequent observations of that species within the hour are ignored. If a species was recorded in the first ten-minute interval, it is allocated a score of 6, the second a score of 5, the third a score of 4, and so on. Unrecorded species are scored as 0. The one-hour count is then repeated, e.g. 10–15 times, and a mean score across all one-hour counts is calculated for each species (Pomeroy & Tengecho 1986).

The method can become increasingly more complex by recording birds within set distance bands (e.g. up to and beyond 25 m) or within set height bands (e.g. above and below 3 m).

Advantages and disadvantages

This method has the advantage of being quick. Common species are ignored once first seen, and thus effort can be concentrated on finding less common species. A reasonably large area can be covered in the allocated time, thus increasing the chances (compared to point counts) of obtaining a complete species list for the site. The method is also easy and does not require prior mapping or cutting of transect lines, and is a good way of rapidly evaluating the importance of sites. If, however, densities of common species are of interest then this is not a suitable method. Timed species counts only provide crude, relative indices of abundance.

Biases

Species vary greatly in their detectability, and thus comparisons between species need to be interpreted cautiously, as do comparisons within species between different habitats. Flocking species or those that aggregate (e.g. in fruiting trees) will have lower indices than those that are dispersed more widely across the study area, even though they may be equally abundant.

Vocal individuality

Rare species that are difficult to see or capture

Method

Many, if not most, species of a wide taxonomic range show individual variations in calls which may be consistent between years and can thus be used as a censusing tool (Saunders & Wooller 1988).

The songs or calls of individuals are tape-recorded, preferably with a directional microphone (Wickstrom 1982), and sound spectrograms produced using readily available computer software. The spectrograms of different individuals can then be visually separated either by a panel of observers or by the much more time-consuming, but more rigorous, approach of measuring the duration and frequency of each component of the spectrogram, and using discriminant function analysis to distinguish between individuals (Gilbert *et al.* 1994).

This technique is by no means foolproof. Despite numerous recordings and measurements of the spectrograms it is not always possible to distinguish between individuals. It is most likely to be useful for distinguishing between individuals in small populations, or for monitoring the survival and movements of animals in small populations. For some species the calls of individuals may vary between years, even though they are consistent within years.

Advantage and disadvantages

This is often the only possible method and produces minimal disturbance. However, obtaining high-quality recordings is hard work, especially as rare species are often widely dispersed.

Biases

Only calling or singing birds (mostly males) can be censused by this method. Females, immature males, non-breeding males, and possibly males at low densities may well be missed because they vocalise less frequently.

References

Alerstam, T. (1990). *Bird Migration*. Cambridge University Press, Cambridge.

Alexander, M. & Perrins, C. M. (1980). An estimate of the numbers of shearwaters on the Neck, Skomer, 1978. *Nature in Wales* **17**, 43–46.

Baillie, S. R. (1991). Monitoring terrestrial breeding bird populations. In *Monitoring for Conservation and Ecology*, ed. by F. B. Goldsmith. Chapman and Hall, London.

Baker, K. (1993). *Identification Guide to European Non-passerines*. British Trust for Ornithology, Thetford, UK.

Berthold, P., Fliege, G., Querner, U. & Winkler, H. (1986). The development of songbird populations in central Europe: analysis of trapping data. *Journal für Ornithologie* **127**, 397–437.

Bibby, C. J., Burgess, N. D. & Hill, D. A. (1992). *Bird Census Techniques*. Academic Press, London.

Bibby, C. J., Phillips, B. N. & Seddon, A. J. (1985). Birds of restocked conifer plantations in Wales. *Journal of Applied Ecology* **22**, 619–633.

Birkhead, T. R. & Nettleship, D. N. (1980). Census methods for Murres *Uria* species: a unified approach. *Occasional Papers of the Canadian Wildlife Service*, No. 43.

Blondel, J. & Frochot, B. (eds.) (1987). Bird census and atlas studies. *Acta Oecologica (Oecologia Generalis)* **8**, No. 2.

BTO (1984). *Ringers Manual*. British Trust for Ornithology, Tring.

Buckland, S. T. (1987). On the variable circular plot method of estimating density. *Biometrika* **43**, 363–384.

Buckland, S. T. & Baillie, S. R. (1987). Estimating bird survival rates from organised mist-netting programmes. In: Ringing Recovery Analytical Methods, *Acta Ornithologica* **23**, 89–100.

Buckland, S. T., Anderson, D. R., Burnham, K. P. & Laake, J. L. (1993). *Distance Sampling – Estimating Abundance of Biological Populations*. Chapman & Hall, London.

Bullock, I. D. & Gomersall, C. H. (1981). The breeding population of terns in Orkney and Shetland in 1980. *Bird Study* **28**, 187–200.

Burnham, K. P., Anderson, D. R. & Laake, J. L. (1980). Estimation of density from line transect sampling of biological populations. *Wildlife Monographs* **72**, 1–200.

Catchpole, C. K. (1973). The functions of advertising song in the Sedge Warbler *Acrocephalus schoenobaenus* and the Reed Warbler *A. scirpaceus*. *Behaviour* **46**, 300–320.

Cayford, J. T. & Walker, F. (1991). Counts of male Black Grouse *Tetrao tetrix* in north Wales. *Bird Study* **38**, 80–86.

Clobert, J., Lebreton, J. D. & Allaine, D. (1987). A general approach to survival estimation by recaptures or resightings of marked birds. *Ardea* **75**, 133–142.

Dawson, D. K. (1985). A review of methods for estimating bird numbers. In *Bird Census and Atlas Studies*, ed. by K. Taylor, R. J. Fuller & P. C. Lock, pp. 27–33. British Trust for Ornithology, Tring.

DeSante, D. F., Burton, K. M. & Williams, O. E. (1993). The Monitoring Avian Productivity and Survivorship (MAPS) program second (1992) annual report. *Bird Populations* **1**, 1–28.

Eastwood, E. (1967). *Radar Ornithology*. Methuen, London.

Emlen, J. T. (1977). Estimating breeding season bird densities from transect counts. *Auk* **94**, 445–468.

Enemar, A. (1959). On the determination of the size and composition of a passerine bird population during the breeding season. *Vår Fågelvärld*, Supplement **2**, 1–114.

Evans, W. R. (1994). Nocturnal flight call of Bicknell's Thrush. *The Wilson Bulletin* **106**, 55–61.

Fuller, R. J. & Langslow, D. R. (1984). Estimating numbers of birds by point counts: how long should counts last? *Bird Study* **31**, 195–202.

Furness, R. W. & Greenwood, J. J. D. (eds.) (1993). *Birds as Monitors of Environmental Change*. Chapman & Hall, London.

Gibbs, J. P. & Wenny, D. G. (1993). Song output as a population estimator: effect of male pairing status. *Journal of Field Ornithology* **64**, 316–322.

Gibbs, J. P., Woodward, S., Hunter, M. L. & Hutchinson, A. E. (1988). Comparison of techniques for censusing Great Blue Heron nests. *Journal of Field Ornithology* **59**, 130–134.

Gilbert, G., McGregor, P. K. & Tyler, G. (1994). Vocal individuality as a census tool: practical considerations illustrated by a study of two rare species. *Journal of Field Ornithology* **65**, 335–348.

Goldsmith, F. B. (ed.) (1991) *Monitoring for Conservation and Ecology*. Chapman & Hall, London.

Green, R. E. & Hirons, M. G. J. (1988). Effects of nest failure and spread of laying on counts of breeding birds. *Ornis Scandinavica* **19**, 76–78.

Hagemeijer, W. & Verstrael, T. (eds.) (1994). *Bird Numbers 1992. Distribution, Monitoring and Ecological Aspects*. SOVON, Beek-Ubbergen.

Haila, Y., Jarvinen, O. & Koskimies, P. (eds.) (1990). Monitoring bird populations in varying environments. *Annales Zoologici Fennici* **26**, 149–330.

Harris, M. P. (1984). *The Puffin*. T. & A. D. Poyser, Calton.

Harris, M. P. & Rothery, P. (1988). Monitoring of Puffin burrows on Dun, St Kilda, 1977–1987. *Bird Study* **35**, 97–99.

Hatch, S. A. & Hatch, M. A. (1989). Attendance patterns of Murres at breeding sites: implications for monitoring. *Journal of Wildlife Management* **53**, 43–493.

Haug, E. A. & Didiuk, A. B. (1993). Use of recorded calls to detect burrowing owls. *Journal of Field Ornithology* **64**, 188–194.

IBCC (1969). Recommendations for an international standard for a mapping method in bird census work. *Bird Study* **16**, 248–255.

James, P. C. & Robertson, H. A. (1985). The use of playback recordings to detect and census nocturnal burrowing seabirds. *Seabird* **7**, 18–20.

Järvinen, O. & Väisänen, R. A. (1975). Estimating relative densities of breeding birds by the line transect method. *Oikos* **26**, 316–322.

Johnsgaard, P. A. (1994). *Arena Birds: Sexual Selection and Behaviour*. Smithsonian Institution Press, Washington DC.

Kelsey, M. G. (1989). A comparison of the song and territorial behaviour of a long distance migrant, the Marsh Warbler *Acrocephalus palustris*, in summer and winter. *Ibis* **131**, 403–414.

Kendeigh, S. C. (1944). Measurement of bird populations. *Ecological Monographs* **14**, 67–106.

Kerlinger, P. (1989). *Flight Strategies of Migrating Hawks*. The University of Chicago Press, Chicago.

Komdeur, J., Bertelsen, J. & Cracknell, G. (1992). Manual for aeroplane and ship surveys of waterfowl and seabirds. *IWRB Special Publication* No. 19. Slimbridge, UK.

Koskimies, P. & Väisänen, R. A. (eds.) (1991). *Monitoring Bird Populations*. Finnish Museum of Natural History, Helsinki.

La Perriere, A. J. & Haugen, A. O. (1972). Some factors influencing calling activity in wild mourning doves. *Journal of Wildlife Management* **36**, 1193–1199.

Lebreton, J., Burnham, K. P., Clobert, J. & Anderson, D. R. (1992). Modelling survival and testing biological hypotheses using marked animals: a unified approach with case studies. *Ecological Monographs* **62**, 67–118.

Lloyd, M. C., Tasker, M. L. & Partridge, K. (1991). *The Status of Seabirds in Britain and Ireland*. Poyser, London.

Lowery, G. H. & Newman, R. J. (1966). A continentwide view of bird migration on four nights in October. *Auk* **83**, 547–586.

Marchant, J. H. (1983). *BTO Common Bird Census Instructions*. British Trust for Ornithology, Tring.

Marchant, J. H. (1994). The new Breeding Bird Survey. *British Birds* **87**, 26–28.

Marchant, J. H., Hudson, R., Carter, S. P. & Whittington, P. (1990). *Population Trends in British Breeding Birds*. British Trust for Ornithology, Tring.

Meyburg, B. U. & Chancellor, R. D. (eds.) (1989). *Raptors in the Modern World*. World Working Group on Birds of Prey and Owls, Berlin.

Meyburg, B. U. & Chancellor, R. D. (eds.) (1994). *Raptor Conservation Today*. Pica Press, for the World Working Group on Birds of Prey and Owls, Berlin.

Meyers, J. M. & Pardieck, K. L. (1993). Evaluation of three elevated mist-net systems for sampling birds. *Journal of Field Ornithology* **64**, 270–277.

Mosher, J. A., Fuller, M. R. & Kopeny, M. (1990). Surveying woodland raptors by broadcast of conspecific vocalisations. *Journal of Field Ornithology* **61**, 453–461.

Munn, C. A. (1993). Tropical canopy netting and shooting lines over tall trees. *Journal of Field Ornithology* **64**, 454–463.

Newton, I. & Chancellor, R. D. (eds.) (1985). Conservation studies on raptors. *ICBP Technical Publication* No. 5.

Ormerod, S. J., Tyler, S. J., Pester, S. J. & Cross, A. V. (1988). Censusing distribution and population of birds along upland rivers using measured ringing effort: a preliminary study. *Ringing and Migration* **9**, 71–82.

Owen, M. (1971). The selection of feeding sites by White-fronted Geese in winter. *Journal of Applied Ecology* **8**, 905–917.

Palmeirim, J. M. & Rabaça, J. E. (1993). A method to analyse and compensate for time-of-day effects on bird counts. *Journal of Field Ornithology* **65**, 17–26.

Pardieck, K. & Waide, R. B. (1992). Mesh size as a factor in avian community studies using mist nets. *Journal of Field Ornithology* **63**, 250–255.

Peach, W. & Baillie, S. R. (1991). Population changes on constant effort sites 1989–1990. *BTO News* **173**. 12–14.

Peach, W. J., Buckland, S. T. & Baillie, S. R. (1990). Estimating survival rates using mark–recapture data from multiple ringing sites. Proceedings of the second EURING Technical Conference, *The Ring* **13**, 87–102.

Pollock, K. H., Nichols, J. D., Brownie, C. & Hines, J. E. (1990). Statistical inference for capture–recapture experiments. *Wildlife Monographs* No. 107.

Pomeroy, D. & Tengecho, B. (1986). Studies of birds in a semi-arid area of Kenya. III – The use of 'timed species counts' for studying regional avifaunas. *Journal of Tropical Ecology* **2**, 231–247.

Porter, R. F. & Beaman, M. A. S. (1985). A resume of raptor migration in Europe and the Middle East. *Conservation Studies on Raptors*, ed. by I. Newton & R. D. Chancellor, pp. 237–242. ICBP Technical Publication No. 5.

Prater, A. J. (1979). Trends in accuracy of counting birds. *Bird Study* **26**, 198–200.

Purroy, F. J. (ed.) (1983). *Bird Census and Mediterranean Landscape* . University of Leon, Leon.

Pyle, P., Howell, S. N. G., Yunick, R. P. & DeSante, D. F. (1987). *Identification Guide to North American Passerines*. Slate Creek Press, Bolinas, California.

Ralph, C. J. & Scott, J. M. (eds.) (1981). Estimating numbers of terrestrial birds. *Studies in Avian Biology* **6**. Cooper Ornithological Society, Las Cruces.

Rapold, C., Kersten, M. & Smith, C. (1985). Errors in large scale shorebird counts. *Ardea* **73**, 13–24.

Rappole, J. H. & Warner, D. (1980). Ecological aspects of migrant bird behaviour in Veracruz, Mexico. In *Migrant Birds in the Neotropics: Ecology, Behavior, Distribution and Conservation*, ed. by A. Keast & E. S. Morton, pp. 353–395. Smithsonian Institution Press, Washington DC.

Rappole, J. H., McShea, W. J., Vega-Rivera, J. (1993). Evaluation of two survey methods in upland avian breeding communities. *Journal of Field Ornithology* **64**, 55–70.

Reynolds, R. T., Scott, J. M. & Nussbaum, R. A. (1980). A variable circular plot method for estimating bird numbers. *Condor* **82**, 309–313.

Robbins, C. S., Bystrak, D. & Geissler, P. H. (1986). *The Breeding Bird Survey: Its First Fifteen Years, 1965—1979*. Resource Publication 157, United States Department of the Interior, Fish and Wildlife Service, Washington, DC.

Robertson, J. G. M. & Skoglund, T. (1985). A method for mapping birds of conservation interest over large areas. In *Bird Census and Atlas Studies*, ed. by K. Taylor, R. J. Fuller & P. C. Lack, pp. 67–72. British Trust for Ornithology, Tring.

Saunders, D. A. & Wooler, R. D. (1988). Consistent individuality of voice in birds as a management tool. *Emu* **88**, 25–32.

Schwartz, P. (1964). The northern water-thrush in Venezuela. *Living Bird* **2**, 169–184.

Smith, N. G. (1985). Dynamics of the transisthmian migration of raptors between Central and South America. In *Conservation Studies on Raptors*, ed. by I. Newton & R. D. Chancellor, pp. 271–290. ICBP Technical Publication No. 5.

Spellerberg, I. F. (1991). *Monitoring Ecological Change*. Cambridge University Press, Cambridge.

Stastny, K. & Bejcek, V. (eds.) (1990). *Bird Census and Atlas Studies. Proceedings of the XIth International Conference on Bird Census and Atlas Work*. Institute of Systematic & Ecological Biology, Prague.

Svensson, L. (1992). *Identification Guide to European Passerines*. Stockholm.

Tasker, M. L., Hope Jones, P., Dixon, T. & Blake, B. F. (1984). Counting seabirds at sea from ships: a review of methods employed and a suggestion for a standardised approach. *Auk* **101**, 567–577.

Taylor, K., Fuller, R. J. & Lack, P. C. (1985). *Bird Census and Atlas Studies*. British Trust for Ornithology, Tring.

Terborgh, J. (1989). *Where Have All the Birds Gone*? Princeton University Press, Princeton.

Thompson, K. R. & Rothery, P. (1991). A census of Black-browed Albatross *Diomeda melanophrys* population on Steeple Jason Island, Falkland Islands. *Biological Conservation* **56**, 39–48.

Tomiałojć, L. (1980). The combined version of the mapping method. In *Bird Census Work and Nature Conservation*, ed. by H. Oelke. Dachverband Deutscher Avifaunisten, Göttingen.

Verner, J. (1985). Assessment of counting techniques. *Current Ornithology*, Vol. 2, ed. by R. F. Johnson, pp. 247–302. Plenum Press, New York.

Walsh, P. M., Halley, D. J., Harris, M. P., del Nevo, A., Sim, I. M. W. & Tasker, M. L. (1995). *Seabird Monitoring Handbook for Britain and Ireland*. JNCC, Peterborough.

Wanless, S. & Harris, M. P. (1984). Effects of date on counts of nests of Herring and Lesser Black-backed Gulls. *Ornis Scandinavica* **15**, 89–94.

Wickstrom, D. C. (1982). Factors to consider in recording avian sounds. In *Acoustic Communication in Birds*, ed. by D. E. Kroodsma & E. H. Miller. Academic Press, New York.

9 Mammals

William J. Sutherland

School of Biological Sciences, University of East Anglia, Norwich NR4 7TJ,
United Kingdom

Introduction

Although some mammals are easily seen, many species are highly secretive, often hidden
from view, for example by being underwater or nocturnal. In some studies of mammals the
study species is very rarely seen and a few have even been carried out without the study
species being seen at all! A number of the mammal species that attract most conservation
interest occur at low densities. Those species that are both secretive and occur at low densities
are extremely difficult to census. As a result of the problems of seeing many species, other
measures are often used, such as the density of dung, footprints, or feeding signs. These
techniques usually give an index of abundance rather than a measure of density.

Many larger mammals can simply be counted. Unfortunately numerous studies have
shown that many individuals are missed and as a result considerable thought needs to be put
into minimising and qualifying such bias.

Total counts

Conspicuous large mammals

Method

Total counts of large mammals can be made by dividing the area into blocks and counting
the number of individuals in each. This thus requires accurate map reading. Different
observers can count the different sections simultaneously. Total counts can also be made
from the air.

Advantages and disadvantages

This method can provide a count of all individuals and information on distribution. Data on
age and sex composition can often be collected at the same time. This is a good method for
conspicuous species in open areas where the count can be completed rapidly relative to the
speed with which animals move between blocks. It is a good method for species that form few
large groups, e.g. Buffalo *Syncerus caffer* (Dublin *et al.* 1990). Aerial counts have some
advantages over line transects, for example the pilot can fly repeatedly over a group of
animals or a patch of scrub to ensure the count is accurate. If the areas are large it can be

260

Table 9.1. *The use of different methods for different groups.*
* method usually applicable, + method often applicable, ? method sometimes applicable. The page number for each method is given. Marsupials and Monotremes should be referred to under their ecological equivalents.

	Carnivores	Sea mammals	Primates	Ungulates	Bats	Rodents	Rabbits, hares & pikas	Insectivores & elephant shrews	Edentates	Page no.
Total counts	+	+		+						260
Counting breeding sites	+	?	?						?	262
Bat roosts and nurseries					*					263
Line transects	?	+	+	*	?	+	+		+	264
Aerial line transects	?	*		*						266
Individual recognition	+	+	+	+	+					268
Counting calls	+	?	+		+	+	+			269
Mapping calls	+	?								270
Trapping	?			?	?	*	+	*		271
Counting dung	+			*	?	+	*		?	273
Feeding signs	+		?		?	+	?		?	274
Counting footprints and runways	+			+		?	?	?	?	275
Hair tubes and hair catchers						*		+		276
Counting seal colonies		+								277

difficult to assess whether individuals have been missed or counted twice and there are often problems in ensuring that all areas are covered and none are double counted (such problems can be reduced by using satellite navigation). Aerial counts for large areas are often prohibitively expensive. This technique is often used in aerial surveys of sites less than 100 km².

Biases

If individuals can move between blocks during the count this can result in error. It may often be difficult to map read accurately, especially from the air, and this can result in considerable errors.

Counting breeding sites

Species which build conspicuous breeding sites or use obvious holes in the ground

Method

Counting breeding sites entails searching areas for all nest sites. Searches must take place systematically, for example by checking the blocks of a grid or systematically following all rivers along a map. It is necessary to determine whether the breeding site is still in use from signs such as digging, fresh bedding, hairs, droppings, or even the scent (especially useful for carnivores). This requires comparing known current and old sites to learn how to distinguish between these. Trained dogs have been used to find Ringed Seal *Phoca hispida* birth lairs (Smith and Stirling 1978). Beaver *Castor canadensis/C. fiber* lodges can even be counted from the air.

If the average number of individuals using each breeding site can be determined independently then this can be converted to a density estimate.

Advantages and disadvantages

This method may be relatively easy for species with obvious sites but unrealistic for most species. Researchers finding lairs may increase predation pressure by visual, acoustic, or olfactory tracks. Species with lairs often move cubs so counts need to be done within a short period.

Biases

Breeding sites can be misidentified; for example, squirrel dreys can be confused with bird nests. If abandoned sites are mistakenly incorporated then this will result in an overestimate. Some communal breeders may use a range of sites. Daytime resting sites can be mistaken for proper breeding sites.

Bat roosts and nurseries
Bats

Methods

The method entails counting all the individuals. For conspicuous species in buildings and caves this is carried out during daylight with a torch – preferably with a deep red filter (e.g. Kodak Wratten Filter No. 29) to reduce disturbance. This has to be carried out systematically, for example by subdividing the site and counting one section at a time, to ensure no areas are missed or counted twice.

For counting bats inside roosts with large numbers of bats, photographing the clusters and then counting individuals can cause less disturbance, and is more accurate, than manually counting the individuals. For large, dense, hibernating clusters, the bats are too tightly packed to allow individual counts. In these cases, surface area covered in connection with an estimate of the packing density can be used to estimate numbers (Thomas & La Val 1988). In practice both of these are difficult to measure.

Bats may also be counted as they leave the roost. If there are altenative exits these are best counted by simultaneous observers; alternatively the same observer may watch different exits on different nights, although this may lead to errors if the bats switch behaviour. A location should be chosen at which individuals are silhouetted against the sky or, better still, a pale background such as a wall. Do not make high-frequency noises (check with bat detector), for example from rustling nylon or polythene. A bat detector (preferably with earphones) can help check that all individuals are counted. Some bats may return before all individuals leave. It is thus necessary to record all returning individuals, otherwise some may be recorded twice. Species occurring together can often be counted separately owing to differences in size, flight pattern, or time of emergence.

For large roosts, videoing the exit holes while bats leave can yield more accurate counts of individuals; as long as a light is not shone directly on the exit, little disturbance is caused to the bats. In the UK, this would need to be licensed by English Nature.

Counting foliage-roosting Pteropodids can be problematic owing to the often large numbers involved (often from thousands to millions of individuals). In this case an average value is obtained from counting individuals in the roost trees, counting individuals which fly away after being disturbed, and counting bats leaving to forage in the evening. Where the population is very large, the number of trees occupied can be counted and trees can be selected at random and the number of bats counted. (Mutere 1980).

Thomas and La Val (1988) review other techniques, including automatic counters.

Advantages and disadvantages

The method has the great advantage of providing a total count. It can also provide data on distribution.

Caves, and especially mines, may sometimes contain hazardous levels of atmospheric gasses. Visiting bat caves in relatively warm, humid areas runs the real risk of contracting the serious fungal disease of the lungs, *Histoplasmosis*. Constantine (1988) reviews such safety aspects.

Hibernating bats in temperate areas in winter use up some of their fat stores if woken, and long visits or repeated visits may reduce their survival. A license is required to enter the roost of any bat species in the UK, and all bat species are protected from disturbance in all European countries.

Biases

For many species it is hard to find all the breeding or resting sites. Within sites, individuals may still be overlooked or hidden. Not all individuals may be present. For example, the number of bats at roosts varies over short periods. Ideally such biases should be quantified.

Counting hibernation sites is only practical for those species which use caves. Some species may use caves infrequently and this can give a considerable underestimate of population size. Contrary to popular belief, bats are mobile during the winter, and counts may often fluctuate depending on external temperature. For those species which hibernate in crevices, the counts in hibernacula can be under-representative since many individuals may be missed.

Counts at summer roosts may underestimate the total population because they can be very difficult to locate. Counts at roosts can also fluctuate widely, and monitoring over several weeks may be necessary; the best time to count is just prior to parturition when most females are likely to be present.

Strip and line transects

Conspicuous species

This entails travelling along a line, recording the individuals on each side. In strip transects a base line is drawn across the survey area such that perpendicular lines from it would cover the entire area (see Figure 9.1). The base line should be aligned so that the transects tend to cross environmental variations, for example a river valley, to reduce the variance between transects. The base line is then subdivided into widths the same as the transect widths: i.e. if each transect is 400 m wide, 20 km will be divided into 50 transects. A number of these transects are selected using random numbers (see Table 2.6) and these are then surveyed. The transects should be drawn on the map to help one navigate.

In practice, transects may only be possible along roads and tracks. An alternative approach is to stratify the area and carry out transects (see Chapter 2, p. 54) in each. These should be located at random.

The number of individuals counted within a given distance covered at a set speed under standard conditions may also be used to give an index of relative abundance. As described in Chapter 2, transect data is converted into density by one of three techniques: estimating the perpendicular distance between the transect and each individual, counting individuals within

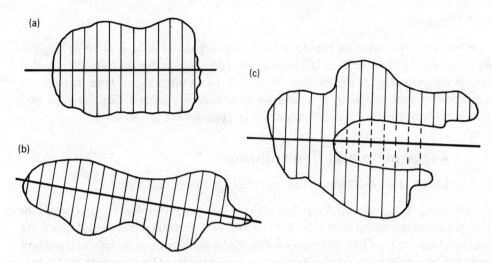

Figure 9.1 Transect designs. In each case the transect is drawn to provide many short transects rather than few long ones. In (c) the dotted areas are not counted.

a set distance, or counting individuals in two bands either nearer to or further away from a set distance. The set distance depends on how open the habitat is and how conspicuous the species is, but a value within 200–400 m is often used.

Transects are often carried out from a vehicle or boat. Walking can be so slow and conspicuous that many individuals that would be seen from a vehicle run away before being counted. Some vehicles have the advantage of raising the observer above the ground. Individuals are more likely to be missed in thick vegetation or when at high densities and thus the observer should slow down.

An indication (although still likely to be an underestimate) of the percentage of individuals visible, yet missed, by the observers can be determined by having two independent teams of observers dictating into separate tracks of a two-track tape-recorder (Marsh & Sinclair 1989). Observer training is very important to increase the chance of detecting individuals.

At night some nocturnal species may be detected by looking for the reflection of light from the back of the retina. Spotlighting can be carried out using a torch, but using a headlight and motorcycle or car battery is much better although much heavier. Spotlighting can often be carried out from a vehicle.

Advantages and disadvantages

This is probably the easiest method for counting species in areas of less than 1000 km^2 of open habitat such as savannah or tundra. It is cheaper than aerial counts and better for small species.

Biases

The method will underestimate the densities for inconspicuous species or those that flee before the observer can count them. It is necessary to consider whether the transect route that is taken is representative of the habitat. Paths and tracks often follow lines that may be unrepresentative, for example by avoiding wet areas and high ground. Sea mammals may often be seen so briefly that it is difficult to judge distance and angle.

Aerial strip and line transects

Large and conspicuous species in open habitat

This entails flying along transects (see p. 264: many of the same principles apply) recording the number of individuals along each side. A 1 : 100 000 or 1 : 50 000 map is needed unless the transects are long. Strips of 100–500 m are used depending on how open the habitat is and how conspicuous the species is. Narrow strips sample a smaller area but fewer animals are likely to be missed. Aim for 10–50 transects. Maps annotated with features make them easier to use in the future. Data is best collected using a tape-recorder. Norton-Griffiths (1978) gives a detailed review of this method and considerable advice for overcoming the usual practical problems.

Only aircraft with high wings give an unobstructed downwards view. If two observers can be carried, double the width can be surveyed. Survey work is normally carried out at 110–190 km/h. It is usual to attempt to count all the individuals within a set distance rather than attempt more complex measures. The edge of the transect can be determined by reference to tapes attached to the window or wing strut (see Figure 9.2). The aircraft flies once down each transect at a height of usually 100 m. In very open country, greater heights can be used, and in dense habitats lesser. The height obviously determines the area censused. Either the correct average height must be maintained or the height regularly recorded and corrected for. In East Africa censuses usually take place at 160 km/h, 100 m height, with a maximum of a 150-m-wide transect; 130 m may, however, be a more sensible minimum safe height (Norton-Griffiths 1978).

Large groups of over 20 animals should be photographed and the photograph number recorded, but the number of animals should still be estimated in case the photographs are unsatisfactory. The recording must specify if the same group is photographed repeatedly, or if overlapping photographs of a group are taken. Photography is less susceptible to individual counting bias especially when previous photographs can be compared and recounted by the same observer.

Training in counting from the air can take place by showing slides from the correct height and angle, asking for an estimate after ten seconds, revealing the correct answer, and then moving on to the next slide. Small groups (e.g. 10) can be counted and the number of groups of that size determined.

Counting efficiency usually declines after three hours. The directional light in early morning and late afternoon should be avoided. If midday is hot, the animals will be sheltering and thus that period should also be avoided.

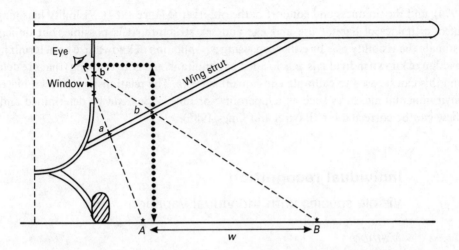

Figure 9.2 How to fix the streamers in position: let h be the height of the observer's eye from the floor, W be the required strip width, and H be the required flying height, then

$$w = W/h$$

w is marked out on the hanger floor, and the two lines of sight, $a'-a-A$ and $b'-b-B$ established. The streamers are attached to the struts at a and b; a' and b' are the window marks. From Norton-Griffiths (1978).

For whales and seals this only gives a relative index unless it is possible to corect for the time spent under the surface. This correction can sometimes be estimated from detailed studies of individuals. Counting the blows of whales and converting that to density removes the problem of estimating group size.

Advantages and disadvantages

This method is expensive and requires careful planning. It is probably the best method for surveying large areas and is often used for large ungulates. Many species that are difficult to count from the ground, for example those that avoid vehicles, can be much easier to count from the air. This is also useful in areas where access is difficult.

Biases

Many individuals will be missed: for example a review of 17 tests of the accuracy of aerial counts of large terrestrial mammals showed that between only 23% and 89% of individuals were seen (Caughley 1974). The proportion seen declines with flight speed and the number of animals. The proportion of individuals seen may vary markedly according to the observer's experience and the visibility (e.g. snow conditions or sea condition) (LeResche & Rausch

1974), and the tiredness and comfort of the observer (Mence 1969). Visibility bias may also affect estimates of habitat use and age and sex structure. It is possible, but difficult, to estimate the visibility bias by either censusing a population of known size or determining the fraction of known individuals (e.g. located by simultaneous radio tracking) that are detected, and this can be used to calibrate subsequent surveys. The relationship between biases and environmental measures such as temperature or time of day can be determined and then these can be corrected for (Bayliss and Giles 1985).

Individual recognition

Visible species with individual variation

Method

This method involves creating a catalogue of all the individuals and their identification features. Photographs are very useful but can be very time-consuming to obtain. At Kruger Park, South Africa, tourists are invited to submit photos of Wild Dogs *Lycaon pictus* (Maddock & Mills 1994). One approach is to provide a record card for each individual with an outline of the species on which identification features may be drawn and any changes are noted. In some species only part of the body may be used, for example the tails of Humpback Whales *Megaptera novaeangliae* (the extent of white on the underside of the tail flukes and shape of the trailing margin), the pattern of spots associated with the whiskers of Lions *Panthera leo* or the form and location of callosities on the heads of Right Whales *Balaena* sp. For other species it is necessary to use a range of characters, for example, for Chimpanzees *Pan troglodytes* the characters used include whether any digits are missing, size, colour, and face shape.

Playback of hyenas in dense habitats has been used with great effect so that they can be seen and identified. Linking trip wires along a path to cameras so that Tigers *Panthera tigris* photograph themselves has been successful.

Advantages and disadvantages

Data on movement and population dynamics can be collected at the same time. Members of the public often show great interest in the history of known individuals and are often prepared to provide sponsorship for named individuals. This method usually provides the best population estimate for carnivores.

This is very time-consuming and is most successful if each individual can be photographed. Hence it is usually of limited use for short-term surveys. It requires good record-keeping, project continuity, and a long-lived target species.

Taking photographs and reconstructing identifications later allows the use of several assistants varying in experience. Good-quality optics and clear views are necessary.

Biases

Individuals may have similar markings and thus be confused. Individuals may obtain new features, such as scars or missing fingers, or scars may heal over and then be categorised as new individuals, unless the study is carried out carefully. Some individuals may be much more approachable than others.

Counting calls
Particularly bats, whales, and seals, but also other species

Method

The usual method is to count the number of calls or bursts of calls and use this as an index of abundance. The vocalisations of whales and seals can be recorded by dangling a hydrophone in the water from a boat, headland, ice edge, or hole in the ice. In an emergency a standard microphone inside an inverted condom can be used as an alternative to a hydrophone. Stirling *et al.* (1983) considered a ten-minute recording long enough to determine the amount and type of vocalisations in their study of wintering arctic pinnipeds. They also had some evidence that some Bearded Seal *Erignathus barbatus* calls could be detected 45 km away. With experience it can be possible in some species, e.g. Humpback Whales, to recognise some individuals from their calls.

Bat detectors work by converting ultrasonic echolocation calls of bats to calls that humans can hear. Detectors can be connected directly to tape-recorder. Earphones or a headset can be used which also makes the censuser less conspicuous to passing humans. The standard technique is to tune the bat detector to a set frequency (45 kHz in the UK) and to record the location and number of contacts or 'bat passes'. Bats in the tropics tend to use higher frequencies. Counting social calls made by bats during song flights have been used as a census method (Hollander 1991, Verheggen 1994). Bats often have regular beats which are patrolled every few minutes, or call from trees (Raaijmakers 1994). This localises the animal, and plotting of the territories can give a good estimation of the breeding male population size.

Bat detectors can also be used for small mammals such as shrews. Counts of long-distance calls (whoops in Spotted Hyenas and roars in Lions) have been used in the Serengeti by calibrating from populations with a known number of individuals (East & Hofer, unpublished).

Advantages and disadvantages

Although these methods are subject to many biases this is often the best method for assessing the relative abundance of these mammals. As mentioned previously, a license is required to enter the roost of any bat species in the UK, and all bat species are protected from disturbance in all European countries.

Biases

Some species can be heard from much greater distances than others (Downes 1982, Waters & Walsh 1994). The calling frequency and the ease with which vocalisations can be detected will vary with environmental conditions, e.g. higher humidity gives lower detectability. The number of calls heard varies considerably with the design of bat detectors or hydrophone, so this must be standardised. Different makes of detector and hydrophone may differ in their sensitivity to the calls of different species (Forbes & Newhook 1990, Stirling *et al.* 1983). Differentiating individuals can be problematic where a single individual calling repeatedly may not be distinguishable from a number of individuals calling less often. Species recognition may also not be possible since many species of bats use calls which are indistinguishable on a bat detector. In this case recording the call and later using sonographic analysis may be the only way to establish its identity. For bats the overlap between the frequency range that can be detected and the frequencies produced is obviously crucial. One study showed considerable differences between bat detectors of the same brand (Thomas & West 1984). This can be tested beforehand by using critical known frequencies. The area covered can be estimated in advance by playing pure frequencies in the required range from different distances from the hydrophone or bat detector.

Mapping calls

Territorial carnivores and primates

Method

The location of calling territorial individuals or groups is mapped, for example using triangulation with pairs of observers in different locations listening simultaneously and recording the compass direction of each call and either recording the time of each call or, preferably, being in radio contact to ensure the same individual is being mapped. If the area can be gridded, mapping is much easier. If the location of simultaneously calling individuals or groups can be determined, this greatly improves the accuracy. If individuals can be identified by recognisable calls they are much easier to map. Using a sonogram may make this easier.

Tape playback can be used to encourage calling. Some species, e.g. Wolves *Canis lupus*, respond to human imitations.

Advantages and disadvantages

This method can be very successful for censusing species at low densities. As with mapping birds (Chapter 8), it may be necessary to visit repeatedly to ensure that the majority of territories are detected.

Biases

Some species may be quiet. Even vocal species may be quiet in some seasons or weather conditions, or at low density. Detectability may vary with habitat. This only determines the number of territorial individuals or groups.

The same problems occur with identification and differential call apparency, as detailed in 'Counting calls' (p. 269).

Trapping

Many mammals, especially small terrestrial species

Method

Many small mammals readily enter traps. In the usual trap design a lever or treddle is pressed which shuts the door behind. There are a large number of trap designs for different-sized mammals, and many of these can be bought commercially. Bateman (1979) describes many designs for species ranging in size from mice to antelopes. Sherman traps (available from H. B. Sherman Inc., PO Box 20267, Tallahassee, Florida 32316, USA) are simple traps obtainable in many sizes and in a collapsible form: useful sizes include $50 \times 62 \times 165$ mm and $76 \times 89 \times 229$ mm. These are easy to carry and use and are reliable. Small and medium macropods can be caught in a Bromilow trap (Kinnear *et al.* 1988). Longworths traps (available from Penlon Ltd, Radley Road, Abingdon, Oxon, OX14 3PH) are the main traps used in Britain. They are effective but bulky. There are numerous designs for traps and modifications, for example for connecting to a time switch (Bekker 1986) or making ant-proof (Beltyukova & Spassky 1989). Pitfall traps can be used in areas which can be dug. These are sunk in the ground either flush or just below the surface. To prevent the catch escaping, high, steep, shiny sides should be used and a rim can be useful. They should be baited and provided with drain holes and also sheltered if overheating or predation is a risk.

Many small mammals are arboreal. Traps can be tied to thick branches, placed on platforms, or tied to stakes in thick grasslands. For aquatic species the traps can be placed on rafts in the water which are then anchored. It is necessary to ensure that the platform will still float once a mammal has been caught and that the anchoring will withstand storms or floods. Alternatively traps can be placed along the water's edge, but the trap must be positioned to ensure that water will not enter even after rain.

Traps are usually placed in pairs with the edge of the trap flush with the ground. Traps should be marked (e.g. with tape in the vegetation above) so they can be re-found. In dense habitats a nylon line can be used to mark the trap line with tape used to mark the position of each trap ('festooning'). Knowledge of the behaviour of the species will greatly improve the trapping success. Most small mammals tend to run alongside logs or other objects rather than across open areas. Placing traps in runs in the vegetation is very successful. A higher density of traps should be used if over 60% are occupied (Gurnell & Flowerdew 1990). Even if few mammals are caught, not more than 200–300 traps can be checked in a day.

On cold nights it will be particularly cold within an exposed metal trap. Nesting material should be provided to ensure that the animal stays warm. Newspaper, straw, dried leaves, or non-absorbent cotton wool are all good. Damp bedding needs to be replaced. Either do not trap in cold weather, visit regularly, or ensure that the traps are well insulated. Polystyrene sheets can be placed under the trap or even completely surrounding it.

Food is usually provided both to increase the catch and sustain caught individuals. Food can be added outside and in the entrance but if too much is provided then animals will not enter the trap. Food is easiest if it is dry and will not decay (e.g. seeds). Peanut butter is a very good bait for rodents and can be mixed with porridge oats and water to produce a thick mix. It is messy and so can be spread on brown bread. For piscivorous species, fresh or tinned fish (not with tomato sauce) can be used but the peanut butter mixture can also work. For small rodents the amount of food provided should be about a tenth of the weight of the largest individual that might be caught. Shrews eat their own weight every 24 hours. Bait should be replenished each day if some has been eaten, it has got wet, or if it may decay.

Prebating is a technique in which the traps are locked open and baited 1–7 days before trapping, since mammals are often neophobic, which results in an increased capture rate once trapping starts. As prebating may attract animals into the study area, this may result in an overestimate of density. Ground baiting entails adding food to the area around the trap locations before trapping. This may increase catching even more than prebating and so is even more likely to distort the distribution. Gurnell (1980) suggests that prebaiting has little effect and any extra days are better spent trapping.

Traps are usually visited twice a day – at dawn and dusk. If shrews may be caught, the traps should be checked every four hours. Small mammal traps can be placed inside a large polythene bag before opening to make escapes less likely.

Shrews are likely to die in traps unless visited regularly (every four hours) and given abundant live food (fly castors bought from fishing shops are good). If shrews are likely to be caught, they need to be catered for even if they are not the subject of the study. Using relatively insensitive traps that shrews will not trigger or adjusting the sensitivity of the trap is one solution.

Population size can be estimated using mark–release–recapture (Chapter 2) (Montgomery 1987). A square grid is usually used, since this minimises edge effects. The spacing will depend on the distance individuals move, but spacing 5–20 m apart is a typical range for small mammals. Each trap should be marked with a cane. Individuals can be marked by tattooing or clipping the guard hairs to reveal the hair underneath, which is often a different colour.

The number of captures divided by the number of trapping nights may also give a reasonable index of abundance (Erlinge 1983). The trap density has to be consistent if this is to be used for comparisons between sites or over time. Half a trap night is subtracted for each trap sprung either accidentally or by catching a mammal.

If the trapping is part of an eradication programme of pests, then depletion methods (Chapter 2) can give useful measures of the initial density.

Mammals that can be seen from a distance may be marked and then resighted as a replacement for recapture. This has been carried out for a number of species of ungulate, often using an aerial survey for the resighting (Bear *et al.* 1989).

Barnett & Dutton (1995) describe the methodology of trapping small mammals in detail.

Advantages and disadvantages

Trapping is very time-consuming, especially for species that occur at low densities, but other data can be obtained while handling animals, for example on sex ratios, weight, or reproductive state. Handling mammals may involve the risk of contracting diseases such as rabies (from bats, rodents, and carnivores – immunisation is required if this is a risk) or Weil's disease (from rodents). If animals are held for a long period while breeding, this may result in abortion or desertion of the young. There are many biases associated with each trap type, size, or means of operating it. Licences are sometimes required, for example to catch shrews in the United Kingdom.

Larger mammals are usually harder to catch and usually occur at much lower densities. Mark–resighting of larger mammals can be costly, especially if aerial surveys are included. Checks are needed to see how often marked individuals are classified as unmarked, leading to a population overestimate.

Biases

Small mammals may be considerably more active on nights when it is dark (Lockhart & Owings 1974), warm (Vickery & Bider 1978), and dry (Mystkowska & Sidorowicz 1961). Some individuals readily enter traps while others evade them.

Counting dung

Many terrestrial and semi-aquatic mammals

Method

For many species their dung is far more conspicuous than the animal itself. If it is possible to identify the species, this is a very good method for detecting the presence of species (Wood 1988).

Simply counting the dung present is one approach. If abundant, then quadrats (see Chapter 3) can be used, but otherwise line transects are suitable. Dung is often highly aggregated and its distribution has been shown to fit a negative binomial distribution. White & Eberhardt (1980) describe a technique for comparing dung counts that differ in their contagion. However, the length of time dung persists may vary between habitats and between time periods, depending upon the weather (especially rainfall), dung composition, fibre content, and the number of dung beetles. There are two ways of overcoming this problem: (1) clear the dung from marked areas, and return before any fresh dung has had time to decay, and count the dung deposited in the interval; (2) mark some very fresh dung (for example with a bamboo stick) and on the revisit use this to compare with dung in the field and ignore any dung that looks more decayed.

Counting dung provides a useful measure of the relative density in different areas. By assessing the amount of dung produced per day from either field observations (remembering that this will vary through the day) or studies of captive individuals, it is possible to convert dropping densities to the number of mammal-days. Captive animals may, however, be fed on a higher-quality diet and may exert less energy and thus may produce an underestimate of the rate at which droppings are produced.

The dung produced by different ages and sexes may differ in size and shape, although there may be overlap (Maccracken & Ballenberghe 1987) and thus there is some possibility for separating the census into such categories.

Advantages and disadvantages

This is often the only practical method for many elusive species. Many other means of censusing large mammals have the problem that visibility varies markedly between habitats. The rate at which dung disappears can be quantified with this method.

Related species may have very similar dung which may either prevent this technique being used or mean it is necessary to survey a combination of species.

Biases

The persistence of dung may vary between habitats and seasons. Defecation rates may vary with diet and season (Rogers 1987) and according to the age and sex of the animal.

Feeding signs

Species with characteristic feeding signs. Particularly useful for herbivores

Method

Many species leave characteristic marks on the remains of their food, for example those which feed on nuts and leave characteristic tooth marks. This is useful for giving data on presence and distribution, and may also be quantified by counting the food density and the proportion that is eaten. It may be necessary to combine species with similar signs, e.g. voles. Surveys can be carried out either within quadrats or line transects. To assess rodent grazing Hansson (1979) determined the proportion of 0.2-m^2 plots with signs of grazing. Bite wounds on livestock have even been used as an index of Vampire Bat *Desmodus rotundus* abundance (Turner 1975).

The method only gives a relative index but can provide a quick and easy measurement which can correlate well with the numbers trapped (Hansson 1979).

It is necessary to determine the food plant species of the study species and how to distinguish the feeding signs from those of other species.

Whether a particular food species is eaten will also depend upon the abundance of other foods.

Counting footprints and runways
Ground-living species

Method

Looking for footprints in areas of soft ground such as near water is a useful way of detecting the presence of species, and counting the density gives a crude but quick indication of abundance. Counts of deer tracks per 16 km of dirt roads correlated reasonably with other estimates of density (Mooty & Karns 1984).

Searching for tracks on soft ground has the disadvantages that soft ground will not be randomly or representatively distributed and it is difficult to tell how old the prints are. A better approach is to standardise by counting the number of individuals crossing a given area. This can be detected by raking areas of ground. On each visit the number of footprints are counted and the area is raked clear to ensure that the same prints are not counted again. For small mammals a tray of dust, sand, or talc can be used but this needs to be protected from rainfall, for example using a canopy. A layer of sand 1 cm thick is ideal. If thicker, tracks of small mammals lose their structure. Fine powders readily become damp and need replenishing.

Tracking stations have been created by coating 1-mm-thick aluminium sheeting with soot from a kerosene torch. However, in one study the relative densities of flying squirrels from trapping and track counts in different sites differed considerably (Carey & Witt 1991).

Another technique is to assess the density of runways made by small mammals. This is done by counting the number of separate runways within a quadrant.

A considerable amount of information can be gained during snow cover on the distribution and behaviour of many mammals.

Scent stations have been used to census a range of carnivore species which occur at a very low density (Diefenbach *et al.* 1994). An area of 1 m diameter is raked and sifted with lime ($CaCO_3$) to make footprints easy to detect. In the middle is placed a scented disc which can consist of fermented egg, carrion, urine, or a synthetic combination of fatty acids in a saturated plaster disc (Roughton 1982). Scent stations are usually placed in groups (or short transects) and the groups should ideally be separated by a distance such that it is unlikely that an individual will visit more than one group. The number of groups which have been visited is counted.

Advantages and disadvantages

This method works well for secretive species but only gives a relative measure. Although different-sized individuals can sometimes be recognised, it is usually difficult to tell if a number of sets of prints are the product of one or more individuals. It may sometimes be necessary to combine species if their footprints cannot be distinguished. This is a particular problem when censusing small mammals.

Biases

This is a measure of both the distance walked and the population density. It will thus be biased by differences in behaviour between different seasons and habitats. If areas are raked they may be avoided or favoured.

Hair tubes and hair catchers

Small mammals (tubes) and carnivores (catchers)

Method

Hair tubes consist of a width slightly greater than the study species (Suckling 1978 used 10-cm lengths of 3-cm-diameter PVC) with double-sided sticky tape stuck to the inside top and folded over at each end. Funnels may be fixed to each end to encourage animals to enter (Scotts & Craig 1988). The tube is fixed to the ground, to a post, or in vegetation, or nailed or strapped to a tree trunk. It may be baited (see under Trapping, p. 271). After 1–2 weeks the tape is either removed in the field and transferred to cards or the trap is taken back to the làboratory. The measurement is the proportion of tubes visited. The duration that the tube is left for should be such as to obtain a high visitation rate in the areas of high mammal densities. If left too long all the tubes may be visited. Each tape is then placed under a binocular microscope and the guard hairs identified from a reference collection. A reference collection can be made of the suitably sized mammals in the region. Hairs of European mammals can be identified referring to Teerink (1991), Australian mammals can be identified referring to Brunner & Coman (1974).

Hair is often caught when individuals squeeze underneath barbed wire fences and this can be useful in providing distribution data. A more systematic approach is to use a loop of multistranded stong wire inserted in the ground so that mammals will squeeze underneath. Clumps of finer wire are inserted within the strands. These clumps should have the texture of a wire brush. This technique is useful for determining which species visits a site – e.g. which predators have visited a bird nest (Pasitschniak-Arts and Messier, in press).

Advantages and disadvantages

This is an easy method that does not require the regular visiting needed for trapping and is less invasive. It is a reasonably easy way of studying arboreal species and species at low densities or over a wide area. The method only provides an index. It may be necessary to spend time learning how to identify the hair of the species present. Separating closely related species may require examining the cross section of the hair. Compared with trapping it is much easier to carry out simultaneous sampling of replicates and may capture species that are not captured by trapping. Extracting, identifying, and counting the hairs can be time-consuming and laborious but this can be done months after the fieldwork.

Biases

The biases described under Trapping (p. 271) also apply here. This method cannot distinguish the number of individuals. Some species leave hairs more readily than others and some are more mobile.

Counting seal colonies

Seals

Method

Seals are usually easiest to count when they return to land to give birth. Seal pups are sometimes counted instead of adults as they show less short-term variation. If counted on the ground they may be sprayed with non-toxic waterproof dye to ensure they are not counted twice. They will lose this marking when they moult.

In some species aerial counts (see also Aerial transects, p. 266) of seal pups or adults can be made from photographs. This needs careful planning to ensure that the entire colony is counted. Using stereoscopic pairs of photographs can improve accuracy. Counting the individuals is best done by overlaying a grid and then counting each block systematically. One technique is to position an acetate sheet over both the photograph and grid and then circle each individual with a pen. This both reduces the risk of missing individuals and also means that it is possible to return to previous photographs to check the constancy of the analysis. Photographs should be independently counted to estimate the errors resulting through individuals being missed (Myers & Bowen 1989). Ultraviolet photography may make individuals easier to detect (Lavigne 1976).

Some studies have tried to correct for the fact that not all individuals are present at a given count. If some adults are marked, are individually recognisable, or are being radio tracked, then the probability of these individuals being present for a given count can be estimated. If these individuals are representative then this can act as a correction factor for the entire colony. For seal pups, the ratio of the number born to peak number ashore can be determined and used as a correction factor, although this may vary between locations (Ward *et al.* 1987).

Advantages and disadvantages

This method can also provide information on distribution and breeding success.

Biases

Individuals may be overlooked or hidden and some may have left the site or been born after the survey. Such biases should be quantified if possible. Probably the most common source of error is missing sites completely.

Acknowledgements

I thank Adrian Barnett, John Dutton, Heribert Hofer, Tony Martin, Martin Perrow, Dave Thompson, and Dean Waters for useful comments on the manuscript. Mike Norton-Griffiths allowed me to reproduce Figure 9.2.

References

Barnett, A. & Dutton, J. D. (1995). *Expedition Field Techniques. Small Mammals*, 2nd edn. Expedition Advisory Centre, Royal Geographic Society, London.

Bateman, J. (1979). *Trapping: A Practical Guide*. David & Charles, Newton Abbot.

Bayliss, P. & Giles, J. C. (1985). Factors affecting the visibility of kangaroos counted during aerial surveys. *Journal of Wildlife Management* **49**, 686–692.

Bear, G. D., White, G. C., Carpenter, L. H., Gill, R. B. & Essex, D. J. (1989). Evolution of aerial mark-resighting estimates of elk populations. *Journal of Wildlife Management* **53**, 908–915.

Bekker, J. P. (1986). A time-clock construction, connected with a Longworth trap. *Lutra* **29**, 222–224.

Beltyukova, O. P. & Spassky, Y. V. (1989). An ant-proof trap for smaller mammals. *Zoologicheskii Zhurnal* **68**, 124–125.

Brunner, H. & Coman, B. J. (1974). *The Identification of Mammalian Hair*. Intaka Press, Melbourne.

Carey, A. B. & Witt, J. W. (1991). Track counts as indices of abundance of arboreal rodents. *Journal of Mammology* **72**, 192–193.

Caughley, G. (1974). Bias in aerial survey. *Journal of Wildlife Management* **38**, 921–933.

Constantine, D. G. (1988). Health precautions for bat researchers. In *Ecological and Behavioural Methods for the Study of Bats*, ed. by T. H. Kunz, pp. 491–528. Smithsonian Institution Press, Washington DC.

Diefenbach, D. R., Conray, M. J., Warren, P. J., James, W. E., Baker, L. A. & Hon, T. (1994). A test of the scent-station survey technique for bobcats. *Journal of Wildlife Management* **58**, 10–17.

Downes, C. M. (1982). A comparison of sensitivities of three bat detectors. *Journal of Mammology* **63**, 343–348.

Dublin, H. T., Sinclair, A. R. E., Boutin, S. Anderson, E., Jago, M. & Arcese, P. (1990). Does competion regulate ungulate populations? Further evidence from Serengeti, Tanzania. *Oecologia* **82**, 283–288.

Erlinge, S. (1983). Demography and dynamics of a stoat *Mustela erminea* population in a diverse community of vertebrates. *Journal of Animal Ecology* **52**, 705–726.

Forbes, B. & Newhook, E. M. (1990). A comparison of the performance of three models of bat detectors. *Journal of Mammology* **71**, 108–110.

Gurnell, J. (1980). The effects of prebating live traps on catching woodland rodents. *Acta Theriologica* **25**, 255–264.

Gurnell, J. & Flowerdew, J. R. (1990). *Live Trapping Small Mammals: A Practical Guide*. Occasional Publications of the Mammal Society, The Mammal Society, Reading.

Hansson, L. (1979). Field signs as indicators of vole abundance. *Journal of Zoology* **206**, 273–276.

Hollander, H. (1991). *Narr een Methode voor Monitoring van Territoriale Mannetjes van de Gewone Dwergvleermuis (Pipistrellus pipistrellus)*. Minsistrie van Landbouw, Natuurbeherr en Visserij, Netherlands.

Kinnear, J. E., Bromilow, R. N., Onus, M. L. & Sokolowski, R. E. S. (1988). The Bromilow trap: a new risk-free soft trap for catching small to medium-sized macropods. *Australian Wildlife Research* **15**, 235–237.

Lavinge, D. M. (1976). Counting Harp Seals with ultra violet photography. *Polar Record* **18**, 269–277.

LeResche, R. E. & Rausch, R. A. (1974). Accuracy and precision of aerial moose censussing. *Journal of Wildlife Management* **38**, 175–182.

Lockart, R. B. & Owings, D. H. (1974). Seasonal changes in the activity pattern of *Dipodomys spectabilis*. *Journal of Mammology* **55**, 291–297.

Maccracken, J. G. & Ballenberge, V. van (1987). Age and sex-related differences in faecal pellet dimensions of moose. *Journal of Wildlife Management* **51**, 360–364.

Maddock, A. & Mills, M. (1994). Population characteristics of African Wild Dogs *Lycaon pictus* in the Eastern Transvaal Lowveld, South Africa, as revealed through photographic records. *Biological Conservation* **67**, 57–62.

Marsh, H. & Sinclair, D. F. (1989). Correcting for visibility bias in strip transect aerial surveys of aquatic fauna. *Journal of Wildlife Management* **53**, 1017–1024.

Mence, A. J. (1969). Psychological problems of conducting aerial censuses from light aircraft. *East African Agriculture and Forestry Journal* **34**, 38–43.

Montgomery, W. I. (1987). The application of capture–mark–recapture methods to the enumeration of small mammal populations. *Symposium of the Zoological Society of London* **58**, 25–57.

Mooty, J. J. & Karns, P. D. (1984). The relationship between White-tailed Deer track counts and pellet-group surveys. *Journal of Wildlife Management* **48**, 275–279.

Mutere, F. A. (1980). *Eidolon helvum* revisited. In *Proceedings of the 5th International Bat Research Conference*, ed. by D. E. Wilson & A. L. Gardner. Texas Technical University Press, Lubbock.

Myers, R. A. & Bowen, W. D. (1989). Estimating bias in aerial surveys of Harp Seal production. *Journal of Wildlife Management* **53**, 361–372.

Mystkowska, E. T. & Sidorowicz, J. (1961). Influence of the weather on captures of micromammalia. II. *Insectivora Acta Theriologica* **5**, 263–273.

Norton-Griffiths, M. (1978). *Counting Animals*. 2nd edn. African Wildlife Leadership Foundation, Nairobi. Available from the African Wildlife Foundation, PO Box 48177, Nairobi, Kenya.

Pasitschniak-Arts, M. & Messier, F. (in press). Predator identification at simulated waterfowl nests using inconspicuous haircatchers and wax-filled eggs. *Canadian Journal of Zoology*.

Raaijmakers, H. (1994). Een onderzoek naar roepende mannetjes van de rosse vleermuis. *Nieusbrief, Vleermuiswerkgroep Nederland* No. 18, pp. 1–5.

Rogers, L. L. (1987). Seasonal changes in defecation rates of free ranging White-tailed Deer. *Journal of Wildlife Management* **51**, 330–333.

Roughton, R. D. (1982). A synthetic alternative to fermented egg as a canid attractant. *Journal of Wildlife Management* **46**, 230–234.

Scotts, D. J. & Craig, S. A. (1988). Improved hair-sampling tube for the detection of rare mammals. *Australian Wildlife Research* **15**, 469–472.

Smith, T. G. & Stirling, I. (1978). Variation in the density of Ringed Seal (*Poca hispida*) birth lairs in the Amundsen Gulf, Northwest Territories. *Canadian Journal of Zoology* **53**, 1297–1305.

Stirling, I., Calvert, W. & Cleator, H. (1983). Underwater vocalisations as a tool for studying the distribution and relative abundance of wintering pinnipeds in the high arctic. *Arctic* **36**, 262–274.

Suckling, G. C. (1978). A hair sampling tube for the detection of small mammals in trees. *Australian Wildlife Research* **5**, 249–252.

Teerink, B. J. (1991). *Hair of West-European Mammals*. Cambridge University Press, Cambridge.

Thomas, D. W. & LaVal, R. H. (1988). Survey and census methods. In *Ecological and Behavioural Methods for the Study of Bats*, ed. by T. H. Kunz, pp. 77–89. Smithsonian Institution Press, Washington.

Thomas, D. W. & West, S. D. (1984). On the use of ultrasonic detectors for bat species identification and the calibration of QMC mini bat detectors. *Canadian Journal of Zoology* **62**, 2677–2679.

Turner, D. C. (1975). *The Vampire Bat*. Johns Hopkins Press, Baltimore, Maryland.

Verheggen, L. (1994). Monitoring van territorial vleermuizen in de paartijd. *Nieusbrief, Vleermuiswerkgroep Nederland* No. 17, pp. 5–8.

Vickery, W. L. & Bider, J. R. (1978). The effect of weather on *Sorex cinereus* activity. *Canadian Journal of Zoology* **56**, 291–297.

Ward, A. J., Thompson, D. & Hiby, A. R. (1987). Census techniques for Grey Seal populations. *Symposia of the Zoological Society of London* No. 58, 181–191.

Waters, D. A. & Walsh, A. L. (1994). The influence of bat detector brand on the quantitative estimation of bat activity. *Bioacoustics* **9**, 205–221.

White, G. C. & Eberhardt, L. E. (1980). Statistical analysis of deer and elk pellet group data. *Journal of Wildlife Management* **44**, 121–131.

Wood, D. H. (1988). Estimating rabbit density by counting dung pellets. *Australian Wildlife Research* **15**, 665–671.

10 Environmental variables

Jacquelyn C. Jones and John D. Reynolds

School of Biological Sciences, University of East Anglia, Norwich NR4 7TJ, United Kingdom

Table 10.1. *Contents of chapter 10*

Introduction

The distribution and abundance of all plants and animals are determined to some extent by abiotic features of the environment. Thus, measurement of environmental variables forms an integral and planned part of most ecological field research. The choice of measurement technique is rarely straightforward, since even the simplest variables such as temperature can be measured with an array of methods differing in precision, practicality and expense. Furthermore, many useful protocols tend to be modified and improved as they are passed around informally among researchers, without finding their way into readily accessible literature. Some recent techniques have been kindly made available to us, and we present them here. Most of our emphasis will be on the simpler and cheaper methods, but we will mention the more expensive techniques where appropriate, and point towards references for additional methods. Many ingeniously simple methods have been devised, but since they do not come with an 'owner's manual', they deserve a fuller explanation, which we have tried to present here.

Wind

There are two parameters in wind measurement: direction and speed. Direction can be measured easily with a compass by noting the direction of cloud movement or vegetation bending. Beware, though, that direction and speed can vary greatly with height, so it is important to measure wind at a height appropriate to your study species (e.g. hawks versus snails), and to be consistent in repeated sampling. *Wind vanes* are fairly reliable if they are high enough to avoid eddies created by local obstructions. A good rule of thumb is to place the wind vane at a distance from the nearest obstruction equal to ten times the height of the obstruction. The direction of the wind is recorded as the direction that the wind is coming from, so a southern wind comes from the south.

Wind velocity can be estimated using the description provided by the *Beaufort scale of winds* which ranges from no wind at Beautfort 0 up to Beaufort number 12, which is hurricane force (Table 10.2). Obviously, this method does not give a very precise measure of wind speed.

Hand-held gauges have the advantage in fieldwork of being lightweight, versatile, and small (e.g. 20×5 cm). The device consists of a transparent cylinder which is held vertically with a hole facing into the wind. The wind forces a plastic float to rise inside the cylinder, and its height is read from a scale calibrated to wind speed. These devices may include compasses to indicate wind direction. They are fairly lightweight, inexpensive, and typically accurate to within ± 3 km h^{-1}.

Cup anemometers are most accurate. They consist of three or four metal cups attached to spokes placed horizontally on a stand. When the wind blows into the cups, the spokes rotate at a speed that can be calibrated to obtain wind velocity. Portable digital anemometers are also available which measure wind speed in a range of units.

Rainfall

For a quick, rough estimate of rainfall, a flat-bottomed pan with straight sides can be placed out in an open area for a set time period. At the end of the time, the amount of rain it contains can be measured and converted to mm fallen per unit time. Beware of evaporation, though, which can be a serious problem, especially during the day.

Rain gauges are more precise, measuring as little as 0.1 or 0.05 mm of rainfall. They usually consist of a cylinder 10–20 cm in diameter with a funnel at its base. Rain entering the cylinder is then directed by the funnel into a narrow tube with a calibrated scale. You can make your own gauge if you convert the volume of water collected to area covered by the cylinder. With any rain-collecting device you still need to be careful of evaporation, splashing, wind, and the effects of surrounding vegetation, which can divert water into or away from your instrument.

Snow can also be measured and converted into rainfall. A general rule of thumb is that 10 cm of snow is equivalent to 1 cm of rainfall, but this can vary widely depending on how compact the snow is. To make a more accurate conversion, snow collected in a pan or in a

Table 10.2. *The Beaufort scale for ecologists*

Beaufort number	Name of wind	Observable features	Field ecologists' impression	Velocity (km/hour)
0	Calm	Smoke rises vertically	You're having a good time	<2
1	Light air	Smoke drifts downwind. Wind does not move wind vane	You're still having a good time	2 to 5
2	Light breeze	Wind felt on face; leaves rustle. Vane moved by wind	It's a bit tricky to photograph insects on plants	6 to 12
3	Gentle breeze	Leaves and twigs in constant motion; wind extends light flag	At least there are no biting insects to contend with	13 to 20
4	Moderate breeze	Raises dust and loose paper; small branches are moved	It's hard to keep your notes from flapping	21 to 29
5	Fresh breeze	Small trees in leaf begin to sway; crested wavelets form on inland waters	You prefer to work in sheltered places	30 to 39
6	Strong breeze	Large branches in motion; whistling heard in telegraph wires; umbrellas used with difficulty	Your tripod is blowing over	40 to 50
7	Moderate gale	Whole trees in motion; inconvenience felt in walking against wind	You're doing this for the good of science	51 to 61
8	Fresh gale	Twigs break off trees; progress generally impeded	You're not being paid enough	62 to 74
9	Strong gale	Slight structural damage occurs (chimney pots and slate removed)	You're thinking about where you've parked your vehicle	75 to 87
10	Whole gale	Seldom experienced inland; trees uprooted; considerable structural damage occurs	You're wondering how you'll get home	88 to 101
11	Storm	Very rarely experienced; accompanied by widespread destruction	You're wondering what shape your home is in	102 to 121
12	Hurricane	At sea, visibility is badly affected by foam and spray and the sea surface is completely white	Time to find a new study site	>121

rain gauge (the funnel may be removed for snow collection) should be melted and the depth of the water measured. Watch for snowdrifts at the entrance to the collecting device!

Temperature

This can be measured most simply using a mercury or alcohol-filled *thermometer*. Thermometers with extended probes in wire casings are also available for places which are difficult to reach.

A *min/max thermometer* is useful for recording minimum and maximum temperatures. It is reset after each reading, and is typically used for 24-hour periods. The best results are obtained by keeping thermometers (and *hygrometers*, see below) in a meteorological box with wooden slats (a *Stevenson screen*). This maintains a uniform temperature that matches the air outside, avoiding surface heating by incident radiation. Specifications can be found in any meteorological textbook or catalogue.

Electronic thermometers are also available comprising a battery-powered meter and metal probe which can sample air, liquid, or soil. The accuracy of such thermometers varies with the measurable range and price.

The main advantage of these thermometers is that they sample a wider range of temperatures than standard mercury thermometers and in fieldwork are less fragile. Electronic sensors can also be connected to a data logger (Box 10.1).

Box 10.1. **Data logger**

This device stores information from various sensors, including thermometers and light meters. Specialist logging systems are also available e.g. for measuring dissolved oxygen, salinity, temperature, conductivity, and pH all simultaneously. However, this system is expensive. Sensor readings may be taken instantaneously or they may be recorded at predetermined intervals over a set time period, and may include maximum and minimum values, averages, or more complex calculations. Data can often be downloaded directly to a computer.

A *thermograph* is useful for a continuous sequence of temperatures recorded on graph paper.

Consult any general apparatus catalogue for choices of thermometers, and watch for carrying cases that will withstand fieldwork. Price differences tend to reflect the temperature range to be measured, the accuracy of the instrument, and the substance to be measured (liquid, air, soil). Note that many other instruments such as oxygen, pH, and conductivity meters also give temperature readings.

Conversion between temperature scales:

$$\text{from } °F \text{ to } °C\text{: } C = \frac{5}{9}(F - 32)$$

$$\text{from } °C \text{ to } °F\text{: } F = \frac{9}{5}C + 32$$

Humidity

The simplest instrument for measuring relative humidity is a *hygrometer*. Dial-type hygrometers give direct readings from an arm which rotates by contraction or expansion of an appropriate material (e.g. hair) with changes in humidity. Readings can be recorded continuously on a rotating paper (a *hygrograph*). Hygrometers are sold with tables for converting differences between the readings into a humidity value. The best results are obtained in a meteorological box (a *Stevenson screen*: see Temperature, p. 283).

Increased accuracy is afforded by a *wet/dry-bulb thermometer*. This consists of two adjacent thermometers, one with a cloth around the bulb. The cloth is attached to a wick in a bottle of water, and capillary action keeps the cloth around the bulb wet. In 100% humidity, there will be no evaporation from the cloth, and the two thermometers will therefore give the same reading. As humidity decreases, evaporation will cause the wet thermometer to give a progressively lower reading than the dry one, because of the latent heat of vaporisation of water.

A *whirling hygrometer* is more versatile and accurate. The two thermometers are swung around on a shaft like a rattle, causing maximal evaporation and accurate readings.

Electronic hygrometers are also available either for min/max readings or for continuous readings of both temperature and humidity. Some will give vapour-pressure deficit, which is the difference between the partial pressure of water vapour in the air and the saturation vapour pressure, which is the maximum pressure possible at a given temperature (e.g. Grace 1983). Vapour pressure deficit is critical for water loss in plants and animals.

pH

The pH indicates the acidity or alkalinity of a solution and is a measurement of hydrogen or hydroxyl ion activity. It can be measured by two main methods: indicator paper and electronic probe. These can be used for water or soil.

Indicator paper

Wide-range indicator paper is available that measures the full pH range (logarithmic scale, 1–14). However, measurements are best fine-tuned with narrow-range paper once the general region has been determined. Indicator paper is cheap, quick and convenient, but less precise than electronic methods. Water-testing kits for pH are also available (see p. 309).

Electronic determination

This involves a pH meter and electrode. The electrode is immersed in the solution and the meter reads the pH. The meter has to be calibrated at intervals using buffer solutions that resist change in pH. Buffer tablets to make up solutions of pH 4, 7, and 9 often accompany a pH meter. At least two of these (the ones at the upper and lower range) must be used and the standardisation done with buffers at an equivalent temperature to the samples (usually room temperature).

Meters come in two forms: portable ones for the field and laboratory ones. Most portable and laboratory-based meters are accurate to within about ± 0.01 pH unit, although laboratory meters can be purchased that are accurate to ± 0.001 pH unit. Bear in mind that the electrodes are fragile and must be kept in distilled water when not in use.

Water pH can be measured directly in the field using a portable meter by dipping the electrode directly into the water. Always take pH paper with you to double check your meter; it is advisable to calibrate the meter frequently because some have a tendency to drift. Alternatively, samples may be collected in clean glass bottles, rinsed in the sample water, and brought back to the laboratory for reading. Fill bottles to the top, avoiding air bubbles, because air contains carbon dioxide, and this can alter the pH. You should select a method according to accuracy required, time and equipment available, and number of sites or samples to be measured.

Soil pH can be determined by mixing one part soil by volume with two parts distilled water (pH 7), waiting for about 10 min, and taking a reading using one of the methods described above. If you cannot measure the pH when you collect the soil, seal the sample in a container to prevent it from drying before the reading is taken. Specialised probes such as spear-head electrodes are also available, allowing direct insertion into semi-solids.

Duration of sunshine

Meteorological suppliers sell simple sunshine recorders consisting of a hollow cylinder which can be set at an inclination appropriate to the latitude. These are mounted at an elevation sufficient for an uninterrupted view of the sky. Sunlight passes through narrow slits in the cylinder and exposes a photo-sensitive chart inside. The resultant trace consists of segments whose length is proportional to the duration of sunlight. An alternative design consists of a spherical lens that traces light tracks onto paper.

Slope angles and height above shore

The slope and height of a census site can have a drastic effect on plants and animals. This is most striking in the intertidal zone of seashores, but it applies to any shoreline study site. Suppliers of survey equipment offer various instruments for making quick, precise readings, including computerised models. We will describe two low-tech methods.

An *Abney level* is relatively inexpensive and commonly used. It is a pocket-sized device which is held against the top of a pole or stick (e.g. at comfortable eye level) at the shoreline, and aimed at another pole which is the same length, located at the target height. It measures the angle of elevation, α. This can be converted to height by measuring the distance between the two points (shoreline to target). The height of the target above the shoreline is then given by (distance $\times \sin \alpha$).

For the economically minded, we suggest the following method, which is best used by two people, but can be adapted for use by one person if necessary. Use a funnel to fill a 10–15-m

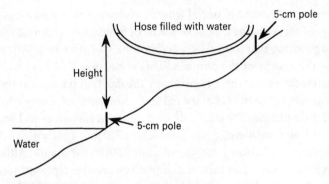

Figure 10.1 Simple method for measuring height above shore.

length of flexible, clear pipe with water from a bucket, and hold your thumbs over each end. One person holds their end of the pipe on the gound at the study site, with the end pointing up about 5 cm above the substrate (Figure 10.1). The other person walks with their end of the pipe down towards the shore, stopping anywhere before the hose is fully stretched out. The shoreward person holds the pipe high enough that when both people release their thumbs, the water stays in the tube. Gravity ensures that when this occurs, the top of the water at each end of the pipe must be at the same height. The down-shore person can now measure the vertical distance from the meniscus to the ground using a tape measure. This gives the vertical drop in elevation between the census site and the point towards the shore. It may be necessary to repeat this procedure, with the up-shore person moving to the down-shore site, and the down-shore person moving farther down until the shoreline is reached. The successive vertical measurements are then added together.

Light

The aspect of light which is measured depends on the biological question being addressed. To choose the best method of measurement, ecologists must therefore understand a little physics, and a lot about what matters to the organisms being considered. Endler (1990) provides an excellent overview of the measurement of light and colour in an ecological context, and Sheehy (1985) provides a more technical discussion. The following is a distillation of those references as well as discussions with colleagues and our own experience.

Photon irradiance

If you are interested in measuring light as available to plants for photosynthesis, or as perceived by animal photoreceptors, you should measure the number of photons striking a unit area. This is the *photon irradiance*, also known as *photon fluence rate* or (almost interchangeably) *photon flux density*. It is typically measured in μmol m^{-2} s^{-1}, where one mole is Avogadro's number of photons ($6.022\,57 \times 10^{23}$). As long as you measure photon flux density within the range of wavelengths which the plants or animals use, then all that matters is the number of photons within that total range, and not the specific wavelength associated with each photon, nor the energy of the photons. This is because a photon has the same photochemical effect irrespective of its energy content – the fact that photons at the blue end of the spectrum have more energy than ones at the red end is irrelevant (cf. Energy flux, below).

Measurements of photon flux density are made with *quantum radiometers*. These usually respond to photons over a range of wavelengths from 400 to 700 nm. This is the visible range, and also the range of photosynthetically active radiation (PAR) for plants. Although it is applicable to most vertebrate photoreceptors, you should beware of exceptions, such as birds seeing ultraviolet (UV). Arthropods typically span 300 nm (UV) to 650 nm. Inferences from your measurements will be only as good as your knowledge of what is meaningful for your particular study species.

Lightweight, portable radiometers are available for field work. The most useful models have a long, rod-like apparatus (e.g. 1 m long) equipped with an array of sensors to integrate over a large area. This accounts for uneven lighting in vegetation; otherwise, one leaf can have a huge effect on the light level you record. A hand-held digital display unit attached to the sensor rod typically stores readings taken over a period of time, to account for the effects of clouds and other temporal changes and the data can be downloaded as point readings or averages.

Energy flux

You may be interested in the energy budget of plants, which includes the radiative, convective, and conductive energy exchanges between plants and their environment (photochemical processes such as photosynthesis involve a very minimal part of the total energy budget). In this case, it is *energy flux* that matters, and this is measured in W m^{-2} (irradiance).

Light energy is measured with *radiometers* (for total radiation) or *spectroradiometers* (for specific wavelengths). Hand-held battery-operated versions are available for field work. *Tube solarimeters* can also be used, though these are more often used in crop stands. They measure energy over the full range of wavelengths from solar radiation. They consist of a glass tube containing a flat theromopile painted with an alternating pattern of black and white rectangles along the midline. The thermopile measures the temperature difference between the black and white rectangles. This information can be recorded by a data logger (Box 10.1), or a millivolt integrator, which is much cheaper and gives accumulated total radiation over a period of time. There are also meteorological standard solarimeters such as the Kipp design, which give an absolute reference because they are cosine corrected (for differing angles of incidence of the light). These are expensive, however, and do not offer the larger area of coverage of tube solarimeters, though they may be useful for standardising your readings for long-term meteorological records.

In the past, equipment for measuring energy was much cheaper than equipment for measuring photons. Thus, energy measurements from devices such as solarimeters were (and still are) used as indirect measurements of light input for photosynthesis. But conversion to photon flux density is not straightforward, and corrections must be made for the different range of wavelength sensitivity of solarimeters. Recent technological advances have improved the portability and price of photon-flux equipment, reducing the need to substitute equipment which measures energy.

Photometers

Standard photographic light meters, often calibrated in *lux*, are useful for studies of human vision, but they have little relevance to anything else. This is because the sensitivity of these meters is based on the spectral responses of 52 pairs of American *Homo sapiens* eyes in 1923! Thus, they are not useful for any animal which does not have the same sensitivity at various wavelengths as humans, and they have no relevance for plants.

Underwater light

Underwater light is commonly measured to determine the depth to which photosynthetic organisms are limited. This occurs when approximately 1% of the surface light still penetrates, and the distance from the surface to this depth is termed the euphotic zone (z_{eu}). This depth can be estimated by examining the visibility of a *Secchi disc* lowered into the water. Readings are best taken under consistent lighting conditions. The Secchi disc is about 30 cm in diameter with alternating black and white or yellow quarters. The disc is lowered slowly into calm water using a calibrated line (for example, marked every half meter) and when the disc disappears, the depth of the line is recorded. Then the disc is lowered slightly further and raised slowly until it reappears. This depth is also recorded. The average of these two depths is the final Secchi-disc visibility reading (Dowdeswell 1984). This reading itself can be used for comparative purposes for a water body over time, but the depth reading (d) has an approximate relationship to euphotic zone depth (z_{eu}), which can be estimated as z_{eu} = between 1.2 to 2.7 times Secchi-disc depth (where the mean is 1.7) (Moss 1988). Clearly, this estimate is very rough and the measured depth will depend on the ambient light, water surface movement, and the person using the disc. If more accurate light readings for a variety of depths are required, then an underwater light meter should be used.

Underwater light meters consist of a probe with a light sensor (measuring photon flux) attached to a recorder or data logger (Box 10.1). Light intensity in the range of photosynthetically active radiation (400–700 nm) is measured by lowering the probe into the water to the appropriate depth. By using selective filters or wavelengths, it can also measure the penetration of narrower ranges of wavelengths (red, blue, and green). This is important because wavelengths are not all absorbed by the water at the same rate; red light is usually absorbed first, followed by green, then blue. Most light meters will measure intensity as photon flux (μmol m^{-2} s^{-1}). Since light is absorbed exponentially with depth, if the irradiance values are logged and plotted against depth, the graph is a straight line. The gradient of this line is known as the extinction coefficient. This value may be used for comparisons between different wavelengths or water bodies. From this graph, the euphotic zone may also be determined.

Water turbidity

Water turbidity, caused by suspended particulate matter, is important in aquatic systems because it reduces the penetration depth of surface light, thereby limiting photosynthesis as well as the visual range of aquatic animals.

The cheapest but most time-consuming method of measuring turbidity is to weigh the particulate matter present in a water sample, which indicates the total suspended solids. If a sample is required from a particular depth, this can be done using a Niskin or Go Flow Flask (General Oceanic Inc.). This consists of a tube with a lid at either end, each attached to a spring closing mechanism. The tube is lowered into the water in the open position and then triggered to trap water at the chosen depth. In the laboratory, a glass-fibre filter paper is rinsed in distilled or deionised water, dried, and then weighed. A standard quantity of the sample is filtered through the dried paper using a Hartley form of a Buchner funnel (Figure 10.2). In this design the base with the holes and the funnel sides are two separate parts held together by clips, so the filter paper covers the entire base, with no gap between the paper and side of the funnel. The used glass-fibre filter is then dried until the weight remains constant, to yield the weight of the particulate matter (original filter weight subtracted from the used filter weight) (Allen 1989).

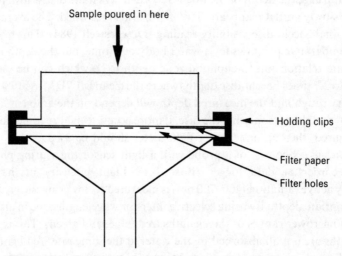

Figure 10.2 Buchner funnel for filtering water samples.

The water sample may also be analysed using a *turbidity meter* (portable or laboratory-based), where some of the sample is put into a sample cell and the light passing through it is measured.

The quickest and most convenient method of measuring turbidity is to use a portable *suspended-solids monitor*. This consists of a meter with a probe containing a fixed light path. The probe is dropped to the depth required, and the turbidity is then read from a meter. The main advantages of this meter are that water samples do not have to be collected, readings are given directly, a variety of pre-set ranges are available and some may take long-term readings if set up to a data logger (Box 10.1).

Water flow

Current velocity can be measured simply by placing a *float* in the water and measuring the time taken to travel a predetermined distance. An orange works well because it does not rise high enough to be affected strongly by wind. This method, though simple and cheap, only measures surface velocity, and gives crude results when there are eddies and variation in velocity within the stream.

A much more accurate method is to use a *flow meter*. This converts the speed of rotation of an impeller to current velocity, and gives readings from specific depths or parts of the stream. Readings should be taken at a variety of distances from shore at 40% of the water column depth. It is customary to do a number of these profiles at intervals along the bank (for example, every 5–10 m) and at set distances from shore. If a more detailed profile is required, readings can be taken at a range of depths, but measurements at a single depth are often satisfactory. Mechanical and electronic versions are available, and these can be hand-held or suspended from a cable.

Conductivity

Conductivity is the measure of the current carried by electrolytes in a solution. In general, seawater has a higher conductivity than freshwater, owing to its higher ionic concentration. Many portable and laboratory-based conductivity meters are available. These may measure either a wide range or a number of narrow ranges which give more accurate readings. Check catalogues for models which suit the accuracy required and your budget.

Salinity

Salinity is the salt content of seawater and is the weight of total salts (g) dissolved in 1 kg of seawater, usually expressed as parts per thousand or ‰. For full strength seawater, this value is usually 35‰. Originally salinity was measured by determining the chlorinity (‰) (which is the mass of chlorine in 1 kg of seawater equivalent to the mass of bromide and iodide) and this is done titrametrically but is not covered here because newer and faster ways to determine salinity have been developed since.

There is a linear relationship between salinity and chlorinity, expressed as (Parsons *et al.* 1984)

$$S\text{‰} = 0.030 + 1.8050 \ Cl\text{‰}$$

Salinity can be measured with a range of methods, either electrically or through simple but fairly reliable gadgets. We will not discuss chemical methods here since they are time-consuming and the methods discussed below are accurate enough for most purposes. Readers interested in chemical methods for salinity and chlorinity determination are referred to Strickland & Parsons (1968).

Conductivity meter

Salinity can be measured by measuring the conductivity of the water (see Conductivity, above) and converting the reading to salinity.

Meters that measure over a single wide range tend to be inaccurate, especially at the freshwater end of the scale. A better option is to use meters that can be focused on whichever part of the scale is relevant, to allow more accurate salinity readings within the subrange.

The conductivity reading in millisiemens (mS) can be used to provide an estimate of salinity (‰) using an equation developed with the aid of Richard Sanders, School of Environmental Sciences, University of East Anglia, Norwich:

$$\text{Salinity} = 0.64 \times \text{Conductivity}$$

This was devised by measuring the conductivity of a dilution series of standard seawater (NaCl) at room temperature. This gives a measurement of salinity which, although not compensated for temperature and pressure, is accurate enough for an estimate of salinity when the major ions are Na and Cl. This method becomes more unreliable as the water becomes less saline. More complicated sets of tables are available for converting conductivity ratio to salinity, with compensation for temperature and depth (NIO 1966).

Salinometer

The salinometer works on a principle similar to the conductivity meter but provides a direct reading of salinity (‰). In some meters the temperature has to be measured separately, and then set on the meter when readings are taken. These machines tend to be more bulky than conductivity meters and come with a range of probes and cable lengths. They are suitable for fieldwork, especially for boat use and for giving readings at depth. Other salinometers are laboratory-based.

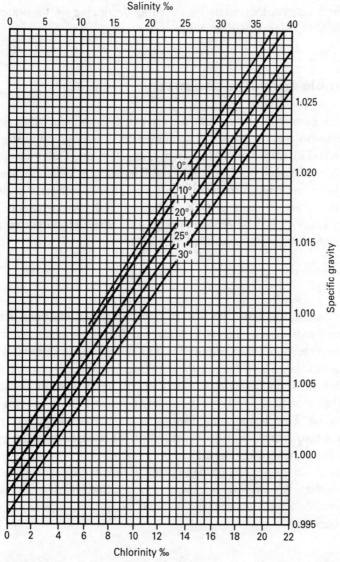

Figure 10.3 Conversion from specific gravity to salinity (after Harvey 1960).

Specific gravity

Salinity can be measured indirectly by a very simple method whereby specific gravity of seawater is recorded with a *hydrometer*. This gives a reading ranging from 1.000 (freshwater) to 1.025 (seawater), which can be converted to salinity when temperature is incorporated. A conversion chart is given in Figure 10.3. A simpler cheap variation of this for use at the seawater end of this range is a device called the *SeaTest Specific Gravity Meter*. This is a small plastic container with an arm that floats to a reading depending on the water's specific gravity. It is available in many aquarium shops.

Preamble to water chemistry

Many substances are present in water which are important for aquatic ecological studies, such as nutrients, silicates, metals, gases, and plant extracts. It would require an entire book to cover all of these, and we recommend Mackereth *et al.* (1978) and Golterman *et al.* (1978) for freshwater, and Strickland & Parsons (1968) and Parsons *et al.* (1984) for seawater. We present methods for quantifying nitrogenous and phosphorus compounds, as they are important to most studies of plant growth, pollution, and eutrophication. First, the following notes on safety should be read, and ecologists who have forgotten their basic chemistry may find the subsequent comments helpful.

Safety

Always be aware of the hazards of the chemicals that you intend to use. Although a symbol is usually given on the container (e.g. corrosive, hazardous, toxic), more detailed hazard data sheets are available on request from the suppliers, and some produce hazard books and CDs providing safety information. These describe the handling, first aid, and disposal procedures that should be followed. Many institutes require that you sign hazard forms which assess your personal risk based on likely levels of exposure. One general rule of safety is to treat all chemicals as hazardous. This means using disposable gloves, both to protect you and to avoid contaminating your samples (especially important for phosphorus). Use goggles and fume cupboards if recommended by hazard forms.

Disposal procedures are normally outlined on the hazard forms. Innocuous chemicals may be flushed down the sink with copious amounts of water. Special disposal procedures for more harmful chemicals and contaminated disposables (gloves, pipette tips, etc.) are available in most institutes.

Glassware

It is important to use inert glassware that will not contaminate your chemicals (e.g. Pyrex). Glassware should be washed beforehand using a harmless surfactant cleaning agent. Sometimes a dilute acid wash is needed to reduce contamination: soak in 10% HCl for 48 hours, then rinse several times in distilled deionised water. This is especially important for nitrogen and phosphorus determination. After washing, store glassware with stoppers or covered with aluminium foil to reduce atmospheric contamination. Some solutions react to light and so should be stored in a dark brown bottle or in a clear bottle wrapped in aluminium foil.

Making up reagents

Solutions are best made up using volumetric flasks, which are available in a wide range of volumes (1 ml to 5 l). It is best to put some liquid into the flask before you add a solid chemical (so it dissolves more easily). Then fill the flask to the measured line, rinsing the sides of the flask and the weighing boat (if used).

The quality of water required depends on the analysis to be done. Water may be available as tap water, distilled water, doubly distilled water, distilled deionised water, or Milli-Q water, or may be purchased as extra pure. For determination of nitrogen and phosphorus, it is advisable to use the water of highest purity available to you (often distilled deionised water) but always check that the water is free from the substance that you want to measure. For dissolved oxygen, either distilled or distilled deionised water is suitable.

In making up standards, anhydrous chemicals must be used. These are obtained by drying them in a drying oven (usually 105°C) overnight then storing them in a desiccator when not in use.

Chemical abbreviations

Weight per volume, w/v: the actual weight of the compound is used instead of the molecular weight. For example, for a 100-ml solution of 10% w/v NaCl, 10 g of NaCl is dissolved in 100 ml of water, as described above.

Volume per volume, v/v: the actual volume of the liquid is used instead of the molecular weight. For example, for a 100-ml solution of 10% v/v HCl, 10 ml of concentrated HCl is added to 90 ml of distilled water.

Dissolved oxygen

Dissolved oxygen (DO) is a critical factor in aquatic ecology. The DO concentration is affected not only by natural factors such as temperature, salinity, plant respiration, and amount of organic material present, but also by organic pollution and eutrophication.

Oxygen in water can be measured using two main techniques: chemically using the *Winkler titration* or electronically with an *oxygen electrode*. Although the chemical titration method is very accurate, it is more time-consuming than using an oxygen meter. Your choice of method will be influenced by the accuracy required and number of samples that you need to analyse. Water testing kits are available to measure dissolved oxygen (see p. 309). There are also *auto-titration machines* which add chemicals automatically, and these can be coupled to home-made units which detect when the titration is complete (the endpoint), although commercial endpoint detectors are now available. Oxygen is expressed either as percentage saturation or in $mg\,l^{-1}$, which is the same as ppm (parts per million).

Winkler titration

Winkler titration in its original form was developed for freshwater but has been modified for many different water types including seawater (Parsons *et al.* 1984), waters rich in organic matter, and waters of high alkalinity (Mackereth *et al.* 1978). The method given in Box 10.2 may be used for both freshwater and seawater samples.

Box 10.2. Winkler titration for measuring dissolved oxygen

The oxygen in a water sample is fixed with addition of the manganous chloride and alkaline iodide solutions. At this stage the sample can be left until it is analysed back in the laboratory. The sample is then acidified in the presence of iodine and then titrated with sodium thiosulphate using a starch indicator. The amount of sodium thiosulphate used in this titration is related to the amount of oxygen in the original sample. The method was modified from Parsons *et al.* (1984).

It is essential to read and follow the safety notes in the Preamble to water chemistry (p. 294) before using this procedure.

Reagents

Manganous chloride
Dissolve 600 g of analytical reagent (AR) grade manganous chloride, $MnCl_2.4H_2O$, in distilled water and make up to 1 litre.

Alkaline iodide
Dissolve 320 g of AR grade sodium hydroxide, NaOH, and 600 g of AR grade sodium iodide, NaI, in distilled water, and leave to cool before making up to 1 litre with distilled water.

Sulphuric acid
Slowly add 280 ml of concentrated sulphuric acid, H_2SO_4, to 500–600 ml of distilled water. Once cooled, make up to 1 litre with distilled water.

Thiosulphate solution
Dissolve 2.9 g of AR grade sodium thiosulphate, $Na_2S_2O_3.5H_2O$, and 0.1 g AR grade sodium carbonate, Na_2CO_3, in distilled water and make up to 1 litre with distilled water. Store in a dark bottle.

Potassium iodate
Dissolve 0.3467 g of anhydrous AR grade potassium iodate, KIO_3, in 200–300 ml distilled water, warming if necessary. Cool and make up to 1 litre with distilled water.

Starch indicator (1% solution)
Dissolve 1 g of starch in 100 ml distilled water by warming to 80–90 °C. This solution should be made up fresh if over 1 week old.

Method

- The water sample is taken using a clean 125-ml reagent bottle and rinsed a few times with the water to be sampled.
- A blank should be prepared in the laboratory using deoxygenated water (purge it with nitrogen) and treated as the sample.
- Fill the bottle using a displacement sampler (Figure 10.4) or long tube reaching into the bottom of the bottle to avoiding 'glugging' and air bubbles which may introduce more oxygen into the sample. The water should be allowed to overflow the bottle before being stoppered.

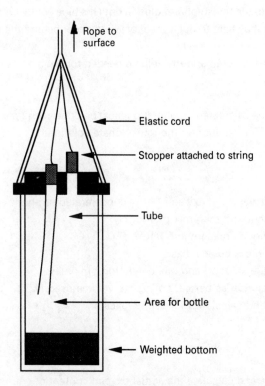

Figure 10.4 Displacement sampler for water oxygen samples. A sampling bottle is placed inside, with the tube inserted into it. The flask is dropped to the required depth and then the rope is jerked. This causes the elastic cord to stretch, pulling out the stoppers and permitting water to flow through the tube and into the bottle.

- Immediately remove the stopper and add 1 ml of the manganous reagent followed by 1 ml of the alkaline iodide solution. In the field this may be added using a 1-ml syringe (one for each solution).
- Stopper the bottle firmly and invert to mix the reagents with the water; a yellow precipitate should form.
- If the sample has been taken in the field or is not to be analysed immediately, it can be kept in this state indefinitely by storing the tightly stoppered bottles under water, preferably in dark conditions at ambient temperature.
- Acidify the sample by adding 1 ml of sulphuric acid reagent to the sample.
- Restopper the bottle and mix the sample thoroughly.
- Within 1 hour of the acidification transfer 50 ml of the sample to a conical flask.
- Titrate the 50-ml sample immediately with the thiosulphate reagent from a burette, using a magnetic stirrer in the flask (or while swirling the flask) until the solution becomes faintly yellow.
- Add 0.5 ml of the starch solution and the solution will turn blue.
- Continue titrating cautiously with the thiosulphate solution until the blue colour disappears and the liquid just turns clear. It may help to put a sheet of white paper behind the flask so the colour change can be observed more easily.
- The volume of thiosulphate used is noted and the volume needed for the blank's titration subtracted to give V_{sample}.

Before the oxygen content can be calculated, a calibration must be done each time a set of samples is analysed to account for variation in the thiosulphate solution.

Calibration

- Fill a 125-ml bottle with distilled water and to it add 1 ml of concentrated sulphuric acid and 1 ml of the alkaline iodide reagent and then mix thoroughly.
- Add 1 ml of the manganous chloride reagent and mix again.
- Transfer 50 ml of this solution into a conical flask.
- Add 5 ml of the potassium iodate solution and mix gently for 2 minutes.
- Titrate with the thiosulphate solution as before, noting the volume used (V_{cal}).
- Do this procedure three times and calculate f for each replicate:

$$f = \frac{5.00}{V_{cal}}$$

Take the mean of the three values and calculate the actual oxygen content of your samples (mg l^{-1} or ppm):

$$\text{mg O}_2 \text{ l}^{-1} = 1.6256 \times f \times V_{sample}$$

Oxygen content of water depends on temperature and salinity (as temperature and salinity increase, oxygen content decreases), so if percentage saturation of dissolved oxygen is required, a theoretical DO content in mg l^{-1} must be calculated. This is the oxygen content in mg l^{-1} of a fully saturated sample, adjusted for temperature and salinity (see Table 10.3). The percentage dissolved oxygen (%DO) content is then calculated using

$$\%DO = \frac{\text{Titrated oxygen concentration (as above)}}{\text{Theoretical oxygen concentration}} \times 100$$

Oxygen electrode

An *oxygen electrode* provides a more convenient and time-saving method of measuring dissolved oxygen. Probes can be portable or laboratory-based and usually consist of a meter (sometimes multipurpose) and a sensor. Although a probe provides direct monitoring, it is less accurate than Winkler titration, which will need to be done occasionally to check that the oxygen probe is calibated correctly and gives accurate readings. A detailed overview of oxygen meters is available in Richardson (1981). When taking readings with an oxygen meter, a water flow must be present or the probe must be kept moving since it abstracts oxygen from the water. Electrodes must be kept clean and moist.

Dissolved oxygen meters need to be calibrated frequently. Most modern portable meters can be calibated easily using zero oxygen solutions and saturated air or 100% oxygen solutions (this can be achieved by filling a bottle with water and shaking it for a minute or two). Some expensive meters calibrate themselves automatically to air.

Scientific catalogues offer a wide range of meters. Your choice will depend on price, accuracy needed (some are accurate to 0.1% while others only to 1%), and your specific requirements. Some meters can be used with more than one type of probe. For example, oxygen, pH, and conductivity can be measured using one meter with three separate probes. Percentage dissolved oxygen concentration depends on temperature and salinity and although most meters will automatically measure and compensate for temperature, only some have a salinity (and altitude) compensation function. Readings from meters lacking temperature or salinity compensation functions can be adjusted using Table 10.3.

Nitrogen

Nitrogen (N) is often determined in water because it is essential for plant growth, and may be a limiting nutrient in seawater. If excessive quantities of nitrogen are present, owing to agricultual run-off, silage, sewage, or industrial discharges, eutrophication may result.

In water, nitrogen exists in gaseous, particulate (organic, for example in plants and algae), and soluble (inorganic) forms. Soluble nitrogen exists in many further forms and constantly fluctuates between them as it is oxidised or reduced:

$$\text{Oxidation} \longrightarrow$$

$$NH_4^+ \leftrightarrow NH_3 \leftrightarrow NO_2^- \leftrightarrow NO_3^-$$
(Ammonium ion) (Ammonia) (Nitrite) (Nitrate)

$$\longleftarrow \text{Reduction}$$

Table 10.3. *Solubility of oxygen in water (modified from HMSO 1988)*

Temperature °C	Solubility of oxygen in water mg l^{-1}	Correction to be subtracted for each degree of salinity $^o/_{oo}$
0	14.62	0.0875
1	14.22	0.0843
2	13.83	0.0818
3	13.46	0.0789
4	13.11	0.0760
5	12.77	0.0739
6	12.45	0.0714
7	12.14	0.0693
8	11.84	0.0671
9	11.56	0.0650
10	11.29	0.0632
11	11.03	0.0614
12	10.78	0.0593
13	10.54	0.0582
14	10.31	0.0561
15	10.08	0.0546
16	9.87	0.0532
17	9.66	0.0514
18	9.47	0.0500
19	9.28	0.0489
20	9.09	0.0475
21	8.91	0.0464
22	8.74	0.0453
23	8.58	0.0443
24	8.42	0.0432
25	8.26	0.0421
26	8.11	0.0407
27	7.97	0.0400
28	7.83	0.0389
29	7.69	0.0382
30	7.56	0.0371

Example. At 15°C, the oxygen content of fully saturated water (100%) is 10.08 mg l^{-1} oxygen, with 0.0546 subtracted for each degree of salinity of the sample.

Ammonium ions (NH_4^+), and nitrate (NO_3^-) are most stable in the environment, and these are the main forms available for plant uptake.

There are four main methods of determining the concentrations of nitrogenous compounds in water. The easiest way is to use a sophisticated machine that does the chemical analyses automatically, the *continuous flow autoanalyser* (Box 10.3). A much cheaper and faster approach which is far less precise and constrained to a narrower detection range with a high detection limit, is the use of a *water testing kit* (see p. 309). A middle option, which we present here in detail, involves individual chemical analyses. These are much cheaper than the autoanalyser and fairly accurate, but time-consuming. Finally, ammonium and nitrate probes are available which are used in conjunction with a pH meter (mV reading). A set of standards is made up and plotted to determine the N concentration (in a similar way to the chemical method). Although this is faster than the chemical method, the detection limit of this electrode, especially the nitrate electrode, may be too high to pick up the low concentrations that are common in many waters.

Box 10.3. **Continuous flow autoanalyser**

This is a machine that analyses your samples using spectrophotometry for dissolved nitrogen and dissolved phosphate using methods similar to those done manually. A continuous steady flow of reagents is present into which a small quantity of your sample is introduced. It moves down through the reagents and then the absorbance is read and stored in a computer. These automated analysers yield results which are highly reproducible, although since reagents are made manually or samples may need to be diluted, some human error still exists.

Recipes for determining the concentrations of nitrate and ammonium are presented below. We recommend that readers interested in determining particulate matter or total nitrogen refer to Mackereth *et al.* (1978) or Parsons *et al.* (1984). Although the methods presented here may be used for both fresh and saline waters, since saline water contains lower concentrations of nitrogenous compounds, the concentrations of the standards may need to be reduced to suit their range. Also note that on mixing salt water with reagents (containing distilled water), some opaqueness will result. The solution should be allowed to settle out before reading the absorbance of a sample. If high accuracy is needed for saline water, the method in Parsons *et al.* (1984) is recommended.

Nitrate

In surface freshwater, nitrate-nitrogen (NO_3-N) is found in concentrations from 0 to almost $10\,mg\ NO_3$-N l^{-1}, usually between 0 to $3\,mg\ NO_3$-N l^{-1}. Groundwater concentrations can reach $30\,mg\ NO_3$-N l^{-1} (Allen 1989, Wetzel 1975). In seawater, the nitrate concentration is usually less than $0.5\,mg\ NO_3$-N l^{-1}. However, be aware that although concentrations increase with contaminants, for example sewage effluent, they are also very much affected by time of year (increases in late autumn and winter owing to agricultural run-off) and the geology of the area. If an autoanalyser is not available (Box 10.3), NO_3-N concentation can be determined using the method in Box 10.4.

Box 10.4. **Nitrate (NO_3-N)**

Nitrate is determined by reducing all of the nitrate to nitrite and then determining this nitrite concentration spectrophotometrically. Any nitrite previously present will also be included, so it is the total oxidised nitrogen (nitrate and nitrite) that is measured, but nitrite concentrations are usually minimal. If high accuracy is required, this nitrite content can be determined using the method found in Mackereth *et al.* (1978). Interference may occur from particulate matter, high densities of dissolved organic matter, or sulphide, in which case the reader is referred to Mackereth *et al.* (1978). The method used here for nitrate determination was kindly made available by Professor Brian Moss of the University of Liverpool, modified from Mackereth *et al.* (1978).

It is essential to read and follow the safety notes in the Preamble to water chemistry (p. 294) before using this procedure.

Reagents

Spongy cadmium
Place zinc rods in a solution of 20% w/v analytical reagent (AR) grade cadmium sulphate, $3CdSO_4.8H_2O$, overnight to build up a cadmium deposit on the rods. Then use a spatula to scrape off the deposits and divide them into small particles. Wash the filings for 15 min with a 2% solution (v/v) of hydrochloric acid (HCl) and then rinse them several times with distilled deionised water to remove all the acid. Filings should always be stored under distilled deionised water to avoid cadmium dust occurring. Just before use, repeat the HCl wash (10 min) and distilled deionised water rinse. Cadmium is poisonous, so wear disposable gloves for handling and dispose of all filings and solutions in a safe manner.

Ammonium chloride
Dissolve 2.6 g of AR grade ammonium chloride, NH_4Cl, in 100 ml of distilled deionised water.

Borax
Dissolve 2.1 g of AR grade borax (di-sodium tetraborate, $Na_2B_4O_7.10H_2O$) in 100 ml of distilled deionised water.

Sulphanilamide
Dissolve 1 g of AR grade sulphanilamide, $NH_2.C_6H_4.SO_2.NH_2$, in 100 ml of 10% v/v hydrochloric acid, HCl.

N-1-naphthylethylene diamine dihydrochloride
Disolve 0.1 g of AR grade N-1-naphthylethylene diamine dihydrochloride,

$C_{10}H_7.NH.CH_2.CH_2.NH_2.2HCl$, in 100 ml distilled deionised water. This solution should be stored in a dark bottle and renewed each month.

Hydrochloric acid (2%)
Add 2 ml of concentrated hydrochloric acid, HCl, to about 50 ml of distilled deionised water and then make up to 100 ml.

Stock nitrate standard
Dissolve 0.722 g of anhydrous AR grade potassium nitrate, KNO_3, in 1 litre of distilled deionised water. This solution contains 0.1 mg NO_3-N ml^{-1} or 100 mg NO_3-N l^{-1}.

Nitrate standards
Make new standards for each use to obtain a calibration curve. Dilute your stock nitrate standard (0.1 mg NO_3-N ml^{-1}) with distilled deionised water as specified below to prepare standards with concentrations of:

   1 mg NO_3-N l^{-1} (10 ml of stock made up to 1 l)
   0.8 mg NO_3-N l^{-1} (8 ml of stock made up to 1 l)
   0.6 mg NO_3-N l^{-1} (6 ml of stock made up to 1 l)
   0.4 mg NO_3-N l^{-1} (4 ml of stock made up to 1 l)
   0.2 mg NO_3-N l^{-1} (2 ml of stock made up to 1 l)
   0 mg NO_3-N l^{-1} (use distilled deionised water)

Method

- Filter your water sample as soon as possible (preferably in the field) and keep it cool. Filtering may be done using a Whatman GF/C glass-fibre filter, handling with forceps to avoid contamination, but bear in mind that the paper may retain or release nitrogen. This can be minimised by first rinsing the paper in distilled deionised water, or using cellulose acetate filters.
- Your sample should be analysed as soon after collection as possible. If it must be stored (which is not recommended as its N content may change), use a filtered sample in a full bottle (excluding air) at 1°C.
- Prepare your standards and treat in the same way as your filtered sample.
- Place 10 ml of your sample water into a 30-ml polystyrene bottle with cap. Add 3 ml of your ammonium chloride solution, 1 ml of your borax solution, followed by 0.5–0.6 g of the spongy cadmium. Screw on the cap and shake in a mechanical shaker for 20 min precisely.
- Transfer 7 ml into a 50-ml volumetric flask and add 1 ml of your sulphanilamide reagent. Mix by swirling.
- After 4–6 min, add 1 ml of your N-1-naphthylethylene diamine dihydrochloride reagent and mix again. Make the solution up to 50 ml with distilled deionised water.
- After 10 min (and before 120 min), measure the absorbance of the samples at 543 nm, using a 1 cm spectrophotometer cell. Use your treated 0-mg NO_3-N l^{-1} standard as your blank;

that is, use it to set your zero on the spectrophotometer so the sample reading automtically has the blank absorbance subtracted from it. This saves you from having to subtract the blank reading from each sample reading.

- Using the absorbance values for your standards, plot a calibration curve (concentration of N as mg NO_3-N l^{-1} against absorbance reading). Your standard graph should be linear. If it levels off, then the top solutions that cause this levelling should not be included. Determine the best-fitting line.

- To determine the concentration of N in your sample, use your fitted-line equation to calculate from the sample's absorbance its related N content (mg NO_3-N l^{-1}). If the sample's absorbance does not fall on the calibration curve line, then dilute some of your original sample with distilled deionised water (for example, twofold or tenfold), and repeat the water analysis procedure. You are aiming for a dilution that brings the absorbance onto the calibration curve line. Multiply your calculated N content by the dilution factor.

Ammonium ions

In surface freshwater, ammonium-nitrogen (NH_4-N) is generally found in the range of 0 to 3 mg NH_4-N l^{-1} though for clean waters the range is usually from 0 to 100 μg NH_4-N l^{-1} (Allen 1989). Again, as for nitrate, the level depends on contamination sources, time of year, and the geology of the area. A manual method for determining the ammonium concentration is given in Box 10.5, though an autoanalyser may also be used if available (see Box 10.3).

Box 10.5. **Ammonium (NH_4)**

The concentration of ammonium is measured by reacting the ammonium with phenol and hypochlorite in an alkaline solution to form indophenol blue. The reaction is catalysed by nitroprusside, and the absorbance of the end product is read spectrophotometrically. The method used here for ammonium determination was kindly made available by Professor Brian Moss of the University of Liverpool, modified from Golterman *et al.* (1978). Interference may occur from very calcareous water, in which case consult Golterman *et al.* (1978). This method measures both ammonium and ammonia, but ammonia concentrations present are usually minimal. Also note that phenol is highly toxic, so an alternative method for freshwater is in HMSO (1981) and for seawater in Parsons *et al.* (1984).

It is essential to read and follow the safety notes in the Preamble to water chemistry (p. 294) before using this procedure.

Reagents

Sodium nitroprusside solution
Dissolve 1 g of analytical reagent (AR) grade sodium nitroprusside, $Na_2[Fe(CN)_5NO].2H_2O$, in 200 ml of distilled deionised water. Store in a dark bottle. This solution is stable for one month.

Take care as sodium nitroprusside is toxic.

Phenol solution
Note that phenol is highly toxic, so handle with extreme care (see Preamble to water chemistry (p. 294)).
Dissolve 20 g of AR grade phenol, C_6H_5OH, in 200 ml of 95% ethanol.

Alkaline solution
Dissolve 100 g of AR grade tri-sodium citrate, $Na_3C_6H_5O_7.2H_2O$, and 5 g of sodium hydroxide, NaOH, in 500 ml of distilled deionised water.

Hypochlorite solution
Using 300 ml of 14% w/v available chlorine hypochlorite solution, make this up to 500 ml using distilled deionised water. Make up fresh for each use.

Oxidising solution
Mix 4 parts of the alkaline solution to 1 part of the hypochlorite solution. 5 ml is required for each sample and should be made up fresh before each use.

Stock ammonium standard
Dissolve 0.9433 g of anhydrous AR grade ammonium sulphate, $(NH_4)_2SO_4$, in 1 l of distilled deionised water. This solution contains 200 μg NH_4-N ml^{-1} or 200 mg NH_4-N l^{-1}.

Working ammonium standard
Take 1 ml of the stock ammonium standard and make up fresh to 500 ml with distilled deionised water for each use. This solution contains 0.4 μg NH_4N ml^{-1} or 400 μg NH_4-N l^{-1}.

Ammonium standards
Using the working ammonium standard, make up the following standards fresh for each use using distilled deionised water, scaling the amounts up if needed:

400 μg NH_4-N l^{-1} (50 ml of the working standard)
300 μg NH_4-N l^{-1} (37.5 ml of the working standard made up to 50 ml)
200 μg NH_4-N l^{-1} (50 ml of the working standard made up to 100 ml)
100 μg NH_4-N l^{-1} (50 ml of the 200 μg l^{-1} standard made up to 100 ml)
50 μg NH_4-N l^{-1} (25 ml of the 100 μg l^{-1} standard made up to 50 ml)
0 μg NH_4-N l^{-1} (use distilled deionised water)

Method

- Filter your water sample as soon as possible (preferably in the field) and keep it cool. Filtering may be done with Whatman GF/C glass-fibre filter, handling with forceps to avoid contamination, but bear in mind that the paper may retain or release nitrogen. This can be minimised by first rinsing the paper in distilled deionised water, or using cellulose acetate filters.
- Your sample should be analysed as soon after collection as possible and kept cool. If it must be stored (not recommended because its N content may change), do so using a filtered sample in a full bottle (excluding air) at 1°C.
- Prepare your standards and treat in the same way as your filtered sample.
- Add 50 ml of your filtered sample to 2 ml of the phenol solution and mix.
- Add 2 ml of the sodium nitroprusside solution and mix.
- Finally, add 5 ml of the freshly prepared oxidising solution, mix, and store the sample in the dark until the absorbance is read.
- After 1.5 hours (and before 12 hours), measure the absorbance of the samples at 640 nm, using a 1-cm glass cell. Use your treated 0-μg NH$_4$-N l^{-1} standard as your blank; that is, use it to set your zero on the spectrophotometer so the sample reading automatically has the blank absorbance subtracted from it. This saves you from having to subtract the blank reading from each sample reading.
- Using the absorbance values for your standards, plot a calibration curve (concentration of N as μg NH$_4$-N l^{-1} against absorbance reading). Your standard graph should be linear. If it levels off, then the top solutions that cause this levelling should not be included. Determine the best-fitting line.
- To determine the concentration of N in your sample, use your fitted-line equation to work out from the sample's absorbance its related N content (μg NH$_4$-N l^{-1}). If the sample's absorbance does not fall on the calibration curve line, then dilute some of your original sample with distilled deionised water (for example twofold or tenfold), and repeat the water analysis procedure. You are aiming for a dilution that brings the absorbance onto the calibation curve line. Multiply your calculated N content by the dilution factor.

Phosphorus

Phosphorus (P) is commonly determined in water because it, like nitrogen, is essential for plant growth, and is often limiting in freshwater. Excessive quantities of phosphorus from sewage treatment works, some industrial effluents, and agricultural run-off can cause eutrophication.

Phosphorus is present in several different forms in water. The fractions most commonly measured are particulate phosphorus (organic, for example in plants and algae) and dissolved phosphorus (soluble reactive phosphorus and soluble unreactive phosphorus). We are usually interested in soluble forms because they are available for uptake by water plants.

Most soluble phosphorus is made up of soluble reactive phosphorus which is orthophosphate (PO_4) and reactive compounds that form PO_4 complexes rapidly. Soluble unreactive phosphorus is the rest of the phosphorus that forms these complexes less rapidly. We present only a method for determining soluble reactive phosphorus since this represents the bulk of the dissolved phosphorus and is the most common form of phosphorus pollutant. However, if the trophic status of freshwater is of interest, it is usually described using total phosphorus. This is the sum of both dissolved and particulate fractions which involves an initial acid digestion of the sample. To measure other forms of phosphorus, for example in studies of nutrient cycling, we recommend Mackereth *et al.* (1978) for freshwater samples and Parsons *et al.* (1984) for seawater samples.

As with nitrogenous compounds, various options are available for measuring phosphorus, including an autoanalyser (see Box 10.3), a water testing kit, and chemical methods described below.

Soluble reactive phosphorus

In freshwater, orthophosphate-phosphorus (PO_4-P) can usually be found in the range of 0–$500\,mg\,PO_4$-$P\,l^{-1}$, but in uncontaminated waters it is commonly found in concentrations of less than $50\,\mu g\,PO_4$-$P\,l^{-1}$. Seawater concentrations are smaller, usually less than $30\,\mu g\,PO_4$-P l^{-1}. Again, as with nitrate and ammonium, the concentration will vary with contamination, time of year, which affects the proportion that is in algal cells, and the geology of the surrounding area. A manual method for determining soluble reactive phosphorus is given in Box 10.6.

Box 10.6. Soluble reactive phosphorus (PO_4-P)

Under acidic conditions, phosphate reacts with molybdate to form molybdo-phosphoric acid. This is reduced to a molybdenum blue complex which is intensely coloured, and this is determined spectrophotometrically. Interference can occur from arsenate, in which case refer to Mackereth *et al.* (1978). The method used here for soluble reactive phosphorus determination was kindly made available by Professor Brian Moss of the University of Liverpool, modified from Mackereth *et al.* (1978).

It is essential to read and follow the safety notes in the Preamble to water chemistry (p. 294) before using this procedure.

Reagents

Sulphuric acid
It is essential to wear disposable gloves and eye protection when handling the concentrated sulphuric acid because it is highly corrosive.

Carefully add 140 ml of concentrated sulphuric acid, H_2SO_4, to 900 ml of distilled deionised water.

Ammonium molybdate solution
Dissolve 15 g of analytical reagent (AR) grade ammonium molybdate, $(NH_4)_6Mo_7O_{24}.4H_2O$, in 500 ml of distilled deionised water. Do not use the solution if more than 6 weeks old.

Ascorbic acid solution
Dissolve 5.4 g of AR grade ascorbic acid, $C_6H_8O_6$, in 100 ml of distilled deionised water. Make this solution up fresh for each use.

Potassium antimonyl tartrate solution
Dissolve 0.34 g of AR grade potassium antimonyl tartrate, $KSbO.C_4H_4O_6$, in 250 ml of distilled deionised water. Do not use the solution if more than 6 weeks old.

Mixed reagent
Make this up fresh each time you analyse samples; 5 ml is required for each sample. Mix the above reagents in these proportions and in this order:

> 5 parts sulphuric acid: 2 parts ammonium molybdate solution: 2 parts ascorbic acid solution: 1 part potassium antimonyl tartrate solution.

Stock phosphate standard
Make up a stock solution by dissolving 4.39 g of anhydrous AR grade potassium dihydrogen orthophosphate, KH_2PO_4, in 1 l of distilled deionised water. This standard solution contains 1 g PO_4-P l^{-1} or 1 mg PO_4-P ml^{-1}.

Working phosphate standard
Add 2 ml of the stock phosphate standard and make up fresh to 500 ml with distilled deionised water. This solution contains 4000 μg PO_4-P l^{-1} or 4 μg PO_4-P ml^{-1}.

Phosphate standards
Using the working standard make up fresh the following standards using distilled deionised water:

> 200 μg PO_4-P l^{-1} (15 ml of working standard, made up to 300 ml)
> 150 μg PO_4-P l^{-1} (75 ml of the 200 μg l^{-1} standard made up to 100 ml)
> 100 μg PO_4-P l^{-1} (100 ml of the 200 μg l^{-1} standard made up to 200 ml)
> 50 μg PO_4-P l^{-1} (50 ml of the 100 μg l^{-1} standard made up to 100 ml)
> 0 μg PO_4-P l^{-1} (use distilled deionised water).

Method

- Soak all glassware, usually overnight, in a weak solution of hydrochloric acid, before rinsing thoroughly with distilled deionised water. This is done because many detergents contain phosphorus and could contaminate your sample.

- Filter your water sample as soon as possible (preferably in the field) and keep it cool. Filtering may be done using a Whatman GF/C glass-fibre filter, handling with forceps to avoid contamination, but bear in mind that the paper may retain or release phosphorus. This can be minimised by first rinsing the paper in distilled deionised water, or using cellulose acetate filters.

- Your sample should be analysed as soon after collection as possible and kept cool. If you must store it (not recommended because its P content may change), do so using a filtered sample in a full acid-washed glass bottle (excluding air) at 1°C.

- Prepare your standards and treat in the same way as your filtered sample. Place 50 ml of this filtered sample or standard into a 100-ml conical flask.

- To each flask, add 5 ml of the mixed reagent.

- After 30 minutes (and preferably well before 12 hours), measure the absorbance of the samples at 885 nm, using a 4-cm glass cell. Use your treated $0 \, \mu g \, PO_4$-$P \, l^{-1}$ standard as your blank; that is, use it to set your zero on the spectrophotometer so the sample reading automatically has the blank absorbance subtracted from it. This saves you from having to subtract the blank reading from each sample reading.

- Using the absorbance values for your standards, plot a calibration curve (concentration of P as μg-$P \, l^{-1}$ against absorbance reading). Your standard graph should be linear. If it levels off, then the top solutions that cause this levelling should not be included. Determine the best-fitting line.

- To determine the concentration of P in your sample, use your actual fitted-line equation to work out from the sample's absorbance its related P content ($\mu g \, PO_4$-$P \, l^{-1}$). If the sample's absorbance does not fall on the calibration curve line, then dilute some of your original sample with distilled deionised water (for example twofold or tenfold), and repeat the water analysis procedure. You are aiming for a dilution that brings the absorbance onto the calibration curve line. Multiply your calculated P content by the dilution factor.

Water testing kits

Kits are available which test the above water conditions (dissolved oxygen, ammonium, nitrate, phosphate) as well as a wide range of other characteristics such as hardness, pH, calcium, iron, and silicate. Examples include water testing tablets (Fisons), water quality test kits (Fisons) and water checkits (Merck Ltd). These kits are handy for quick, rough estimates of many water chemistry parameters and may be used to double-check measurements made using other methods.

Most of these testing kits have been developed to measure drinking water, so beware of the range they measure. For example, nitrate and nitrite kits may be set to detect minimum safe levels which are actually higher than the concentrations found in many freshwater environments.

Many quick individual tests are available for some of the other field measurements discussed in this chapter such as pH and dissolved oxygen, but check their accuracy against your requirements. These methods have the benefit of being simple, quick, and cheaper than electric equipment but are not as accurate. Check scientific catalogues to see what is available and for further details.

Soil characteristics

Soil profiles

Soil profiles can be examined by digging a hole with clean sides and using a measuring tape to record the depth of each layer. Samples taken with an *auger* are less destructive and time-consuming, so more samples can be taken to obtain a better representation. The simplest way is to twist a screw-shaped auger a short distance into the ground, noting the maximum penetration. The auger is then pulled out carefully (watch your back as well as the sample!) and laid on a pan so that the soil can be removed in its original vertical profile. Then successive, deeper samples are taken. Augers are available with a variety of heads for penetrating different soil types.

A *gouge auger* or *cylindrical core sampler* is less time-consuming and more accurate because it can remove an entire core at once. A large metal cylinder (e.g. 22 cm long, 4 cm diameter) can be sharpened at one end and twisted into the ground by rotating a cross-bar inserted through the top (Dowdeswell 1984).

Soil surface hardness, or 'penetrability'

This can be important for seed germination, and for animals which dig or scrape the ground for food, including many birds which probe the substrate for invertebrates. The easiest way to measure penetrability, especially for mudflats and other soft soils, is the 'BJPS', or 'Bob James' pointy stick' method. Hold a sharpened stick or rod at arm's length, drop it straight down, and measure how far it penetrates! This is easiest if the stick is marked in centimetres. Be consistent with the dropping height, and expect to use the average of five or more measurements, depending on variation among readings. Remember to account for temporal variation due to soil moisture content.

More technologically minded researchers can use a *penetrometer*. Lightweight pocket versions are available commercially. The following specifications are for a home-made version which we have used (Figure 10.5), but the exact dimensions and materials are not critical. The version illustrated consists of a length of PVC pipe 1 m long and 3.5 cm in diameter. The top of the pipe meets a T-bar of similar diameter and 30 cm long. This T-bar

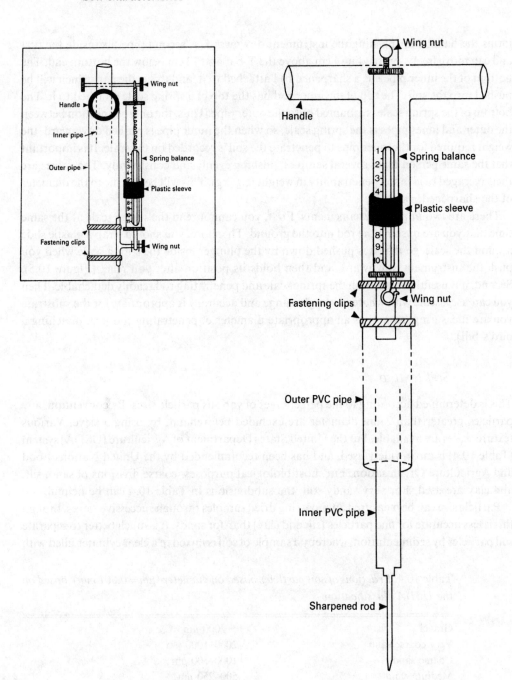

Figure 10.5 Penetrometer, for measuring soil penetrability.

forms the handle, for pushing the instrument downward. A second pipe fits inside the first and can move freely, protruding 1 cm above the T-bar and 13 cm below the bottom end. The bottom of the inner pipe has a sharpened rod attached to it, and this is the part which will be pushed into the soil. The top of the inner rod has the top of a spring scale attached to it. The bottom of the spring scale is mounted onto the outer pipe. Thus, the only connection between the outer and inner pipes is the spring scale, so when the outer pipe is pushed downward, the weight required for the inner pipe to penetrate the soil is recorded on the scale. It is important that the same person takes several samples, pushing evenly and consistently. The values are then averaged to give soil penetrability in weight (e.g. kg). This will be specific to the diameter of the sharpened rod.

There are two important refinements. First, you cannot read the spring scale at the same time that you are pushing the rod into the ground. Therefore, you should wrap a plastic slide around the scale, so that it is pushed down by the plunger inside the spring scale when you push the instrument downward, and then holds its position after you relax (Figure 10.5). Second, it is useful to have both the spring scale and penetrating rod readily detachable. Then you can use a scale with whatever weight range and accuracy is appropriate for the substrate you are measuring, as well as an appropriate diameter of penetrating rod (e.g. matching a bird's bill).

Soil texture

This is determined according to the percentages of various particle sizes. By convention, any particles greater than 2 mm diameter are excluded beforehand, by using a sieve. Various texture scales are available, but the United States Department of Agriculture (USDA) system (Table 10.4) is most widely used, and has been recommended by the United Nations Food and Agriculture Organisation. For most biological purposes, coarse divisions of sand, silt, and clay are used. For very sandy soil, the subdivisions in Table 10.4 can be helpful.

Particle size can be measured after passing dried samples through successive sieves, though this is less accurate for fine particles (silt and clay) than for sands. It is much better to separate soil particles by sedimentation, whereby a sample of soil is mixed in a clear cylinder filled with

Table 10.4. *Fractions of soil particles based on diameter ($\mu m = 0.001$ mm), based on the USDA classification*

Gravel	> 2000 μm
Very coarse sand	2000–1000 μm
Coarse sand	1000–500 μm
Medium sand	500–250 μm
Fine sand	250–125 μm
Very fine sand	125–50 μm
Silt	50–2 μm
Clay	< 2 μm

Box 10.7. **The Bouyoucos method for determining fractions of silt and clay in soil**

The following procedure is from notes kindly provided by Dr David Dent, University of East Anglia.

Place 50 g of soil into a beaker, and add approximately 100 ml of distilled water. Add a dispersing agent such as 10 ml of 10% weight/volume of Calgon. Alternatively, a solution of sodium hydroxide, NaOH, can be used: dissolve 100 g of sodium hydroxide pellets in 200 ml of water, dilute to 2.5 l, and add 10 ml of this to the soil sample. Stir and leave for 15 minutes, make up the volume to 600 ml, and record the temperature. Shake the suspension thoroughly for 1 minute, set the cylinder on the bench, and record the time.

Reading 1: At 30 seconds, insert a *soil hydrometer* into the solution. This is similar to a hydrometer used to measure salinity (p. 293), but is calibrated in g/l. Be careful not to stir up the mixture. At precisely 40 seconds, take a reading to obtain the weight of silt and clay in suspension.

Reading 2: At 6.5 hours take the final reading.

Calculations

To correct for temperature, for each degree above or below 20°C, either add or subtract, respectively, 0.4 g/l. The percentage moisture used in the calculations below is the percentage by weight, obtained as outlined in Soil moisture, below.

$$\text{Percentage sand } (>50\ \mu m) = 100 - \frac{\text{corrected Reading 1} \times 100}{50 - (\text{percentage moisture}/2)}$$

$$\text{Percent clay } (<2\mu m) = \frac{\text{corrected Reading 2} \times 100}{50 - (\text{percentage moisture}/2)}$$

$$\text{Percent silt } (2\text{–}50\ \mu m) = 100 - (\text{percentage sand} + \text{percentage clay})$$

The results are expressed on an oven-dry weight basis, not including organic content (see p. 315).

water. Loose organic material usually floats to the top, and the various particle types settle according to size. The percentages of each fraction can be measured using the *Bouyoucos method* (Box 10.7). The resulting percentages are expressed on a dry weight basis, and can be used to classify soil texture (Figure 10.6). This classification refers to mineral fractions of soil only (see below for organic matter).

Soil moisture

This is broken down for ecological purposes into categories which differ in their availability to plants: (1) free-draining (or gravitational) water drains away soon after rain, but may be available to plants for several days; (2) plant-available (capillary) water occupies the spaces between soil particles, and is most important to plants; (3) plant-unavailable (hygroscopic) water forms a thin film on the surface of particles.

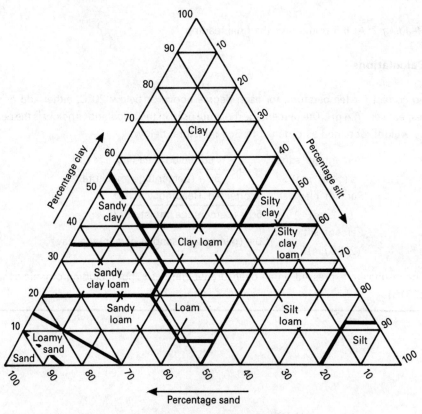

Figure 10.6 Soil texture classification chart (after FAO 1977).

Plant-available water can be approximated by the loss in weight of soil at field capacity (soil that had been saturated and allowed to finish draining (e.g. for two days), when air-dried (e.g. spread out on a newspaper for one week). This method extracts slightly more than the plant-available water, but the difference is not great, except in clay and peat soils. Electronic soil moisture meters are also available for *in situ* field measurements.

Soil organic content

This is important as a source of nutrients to plants and invertebrates. It is measured by drying a sample of soil and examining its percentage weight loss after it has been ignited: (1) oven-dry a sample of 5–10 g overnight, or air-dry until the weight is stable; (2) grind the soil with a pestle and mortar; (3) ignite the sample at 375°C in a furnace for 16 hours; (4) allow to cool in a desiccator; (5) re-weigh the sample. Higher temperatures can be used (e.g. 450°C), but this will result in the loss of carbonates, and structural water will also be lost from clays. If pH tests indicate that the soil is low in carbonates, and clay represents less than approximately 10% of the sample, reasonable results can be obtained if the sample is ignited for 2 hours at 850°C, thereby speeding up the processing time.

Redox potential

The redox or oxidation-reduction potential is a measure of how readily a medium will donate electrons to (reduce) or accept electrons from (oxidise) any reducible or oxidisable substance. Solutions with high redox potentials are highly oxidising. This is important in waterlogged soils, for example, because at low redox potential, nitrate (an important plant nutrient) is reduced to nitrogen, and sulphates are reduced to toxic H_2S.

To measure the redox potential, a platinum electrode is connected to an mV meter. Many pH meters also have an mV scale. It is most convenient to use a combined platinum KCl electrode. Tables are available to relate the oxidation-reduction potential E_h to the ion forms of interest, but redox alone is a useful index of the extent of oxidation or reduction in the system.

Oxygen in soils and sediments

Oxygen content in waterlogged soils or soft sediments may be measured simply, using the following method. If an unglazed, earthenware tube is filled with water, corked, and then placed in the ground, an equilibrium of oxygen between the water and the sediment will be reached. For this to occur the tube should be left for at least a week. Then the tube is recovered and the oxygen content of the water (now equivalent to that in the sand or mud) can be measured by any of the aforementioned methods (Dowdeswell 1984).

Acknowledgements

We appreciate valuable comments on various parts of this chapter by R. R. Boar, A. J. Davy, D. L. Dent, T. D. Jickells, R. Sanders, and D. C. Wildon. We had additional helpful discussions with M. Ausden, C. A. Davenport, J. Farquar, and B. Moss. B. Moss kindly permitted us to use his modified protocols for measuring nitrogen and phosphorus compounds, and D. L. Dent provided his protocol for determining soil particle fractions.

References

Allen, S. E. (1989). *Chemical Analysis of Ecological Materials*. 2nd edn. Blackwell Scientific Publications, Oxford.

Dowdeswell, W. H. (1984). *Ecology: Principles and Practice*. Heinemann Educational Books. Fletcher & Sons, Norwich.

Endler, J. A. (1990). On the measurement and classification of colour in studies of animal colour patterns. *Biological Journal of the Linnean Society* **41**, 315–352.

FAO (1977). *Guidelines for Soil Profile Description*. Food and Agriculture Organisation of the United Nations, Rome.

Golterman, H. L., Clymo, R. S. & Ohnstad, M. A. M. (1978). *Methods For Physical and Chemical Analysis of Fresh Waters*. 2nd edn. Blackwell Scientific Publications, Oxford.

Grace, J. (1983). *Plant–Atmosphere Relationships*. Chapman and Hall, London.

Harvey, H. W. (1960). *The Chemistry and Fertility of Sea Waters*. Cambridge University Press, Cambridge.

HMSO (1981). *Ammonia in Waters* (No. 48). Her Majesty's Stationary Office, London.

HMSO (1988). *Dissolved Oxygen in Waters Amendment*. Her Majesty's Stationary Office, London.

Mackereth, F. J. H., Heron, J. & Talling, J. F. (1978). *Water Analysis: Some Revised Methods for Limnologists*. FBA, Ambleside.

Moss, B. (1988). *Ecology of Fresh Waters*. Blackwell Scientific Publications, Oxford.

NIO (1966). *International Oceanographic Tables*. National Institute of Oceanography of Great Britain and UNESCO.

Parsons, T. R., Maita, Y. & Lalli, C. M. (1984). *A Manual of Chemical and Biological Methods for Seawater Analysis*. Pergamon Press, Oxford.

Richardson, J. (1981). Oxygen meters: some practical considerations. *Journal of Biological Education* **15**, 107–116.

Sheehy, J. E. 1985. Radiation. In *Instrumentation for Environmental Physiology*, ed. by B. Marshall & F. I. Woodward, pp. 5–28. Cambridge University Press, Cambridge.

Strickland, J. D. H. & Parsons, T. R. (1968). *A Practical Handbook of Seawater Analysis*. Fisheries Research Board of Canada, Ottawa.

Wetzel, R. G. (1975). *Limnology*. W. B. Saunders, Philadelphia.

11 The twenty commonest censusing sins

William J. Sutherland

School of Biological Sciences, University of East Anglia, Norwich NR4 7TJ,
United Kingdom

1. NOT SAMPLING RANDOMLY.

    It is very satisfying to sample rarities or rich patches but it ruins the exercise. One common error is just to visit the best sites and use the data to estimate population size.

2. COLLECTING FAR MORE SAMPLES THAN CAN POSSIBLY BE ANALYSED.

    This is a waste of time and may raise ethical and conservation issues.

3. CHANGING THE METHODOLOGY IN MONITORING.

    Unless there is a careful comparison of the different methods, changing the methodology prevents comparisons between years.

4. COUNTING THE SAME INDIVIDUAL IN TWO LOCATIONS AND COUNTING IT AS TWO INDIVIDUALS.

5. NOT KNOWING YOUR SPECIES.

    Knowing your species is essential for considering biases and understanding the data.

6. NOT HAVING CONTROLS IN MANAGEMENT EXPERIMENTS.

    This is the greatest problem in interpreting the consequences of management.

7. NOT STORING INFORMATION WHERE IT CAN BE RETRIEVED IN THE FUTURE.

    The new warden of a national nature reserve in England could find out from old work programmes the days on which his predecessor had counted a rare orchid but could find no record of the actual numbers!

8. NOT GIVING PRECISE INFORMATION AS TO WHERE SAMPLING OCCUR-RED.

    Give date and precise location. 'Site A', 'behind the large tree' or 'near to the road' may be sufficient now but mean nothing later.

9. COUNTING IN ONE OR A FEW LARGE AREAS RATHER THAN A LARGE NUMBER OF SMALL ONES.

    A single count gives no measure of the natural variation and it is then hard to see how significant any changes are. This also applies to quadrats.

10. NOT BEING HONEST ABOUT THE METHODS USED.
    If you only survey butterflies on warm still days or place small mammal traps in the locations most like to be successful then this is fine but say so. Someone else surveying on all days or randomly locating traps may otherwise conclude that the species has declined.

11. BELIEVING THE RESULTS.
    Practically every census has biases and inaccuracies. The secret is to evaluate how much these matter.

12. BELIEVING THAT THE DENSITY OF TRAPPED INDIVIDUALS IS THE SAME AS THE ABSOLUTE DENSITY.

13. NOT THINKING ABOUT HOW YOU WILL ANALYSE YOUR DATA BEFORE COLLECTING IT.

14. ASSUMING YOU KNOW WHERE YOU ARE.
    This can be a problem when marking individuals on maps or when censusing areas, e.g. a one-kilometer square marked on a map. Population overestimates can result from incorrectly marking the same individuals as occupying very different locations or by surveying a larger block than intended.

15. ASSUMING SAMPLING EFFICIENCY IS SIMILAR IN DIFFERENT HABITATS
    Differences in physical structure or vegetation structure will influence almost every censusing technique and thus confound comparisons.

16. THINKING THAT SOMEONE ELSE WILL IDENTIFY ALL YOUR SAMPLES FOR YOU.

17. NOT KNOWING WHY YOU ARE CENSUSING
    Think exactly what the question is and then what data you need to answer it. It is nice to collect additional data but will this slow down the project so that the objectives are not accomplished?

18. DEVIATING FROM TRANSECT ROUTES.
    On one reserve the numbers of Green Hairstreaks *Callophrys rubi* seen on the butterfly-monitoring transect increased markedly one year. It turned out that this was because the temporary warden that year climbed through the hedge to visit the colony on the far side.

19. NOT HAVING A LARGE ENOUGH AREA FOR NUMBERS TO BE MEANINGFUL.
    If it is impossible to have a large enough area then question whether the effort might not be better spent on another project.

20. ASSUMING OTHERS WILL COLLECT DATA IN EXACTLY THE SAME MANNER AND WITH THE SAME ENTHUSIASM.
    The International Biological Programme gave very specific instructions, yet it was hard to make much sense of the data because the slight differences in interpretation led to very different results.

Index

Note: page numbers in *italics* refer to figures and tables